PHARMACOLOGY - RESEARCH, SAFETY TESTING AND REGULATION

PENICILLIN

BIOSYNTHESIS, APPLICATIONS AND ADVERSE EFFECTS

PHARMACOLOGY - RESEARCH, SAFETY TESTING AND REGULATION

Additional books in this series can be found on Nova's website under the Series tab.

Additional E-books in this series can be found on Nova's website under the E-books tab.

PHARMACOLOGY - RESEARCH, SAFETY TESTING AND REGULATION

PENICILLIN

BIOSYNTHESIS, APPLICATIONS AND ADVERSE EFFECTS

GENEVIÈVE BARRESI
AND
PERE KOMULAINEN
EDITORS

Nova Science Publishers, Inc.
New York

For permission to use material from this book please contact us:
Telephone 631-231-7269; Fax 631-231-8175
Web Site: http://www.novapublishers.com

LIBRARY OF CONGRESS CATALOGING-IN-PUBLICATION DATA

Penicillin : biosynthesis, applications, and adverse effects / editors, Genevihve Barresi and Pere Komulainen.
p. ; cm.
Includes bibliographical references.
ISBN 978-1-62100-233-8 (softcover)
I. Barresi, Genevihve. II. Komulainen, Pere.
[DNLM: 1. Penicillins--biosynthesis. 2. Penicillins--therapeutic use. 3. Penicillins--adverse effects. QV 354]
LC classification not assigned
579.5'654--dc23 2011032396

Published by Nova Science Publishers, Inc. † New York

CONTENTS

Preface		**vii**
Chapter 1	Molecular Modification of Biosynthetic Genes for Improved Production of Penicillins and Cephalosporins *K. S. Goo, D. S. H. Tan and T. S. Sim*	**1**
Chapter 2	Developments in the Separation and Recovery of Penicillin G Using Hollow-fiber Membrane Contactors *Monwar Hossain*	**59**
Chapter 3	Novel Peptide Endowed with Antibiotic Properties *Séverine Padiolleau-lefevre,* *Denis Phichith and Bérangère Avalle*	**97**
Chapter 4	Effects of Penicillin on Clinical Tests with Special Reference to Serum Albumin Measured by a Modified BCP Method *Yuzo Kayamori, Miyuki Ono,* *and Dongchon Kang*	**113**
Chapter 5	Biodegradability Improvement of Penicillin Formulation Effluent by Ozonation and the Fenton Reagents *Idil Arslan-Alaton*	**127**

Chapter 6 Geometrical and Electronic Structure of Molecule Benzilpenicillin by Method AB INITIO **145**
V. A. Babkin, V. Yu. Dmitriev and G. E. Zaikov

Index **149**

PREFACE

This book presents current research in the study of the biosynthesis, applications and adverse effects of penicilllin. Topics discussed in this compilation include the molecular modifications of biosynthetic genes for improved production of penicillins and cephalosporins; developments in the separation and recovery of penicillin G using hollow-fiber membrane contractors and the effects of penicillin on clinical tests with special reference to serum albumin measured by a modified BCP method.

Chapter 1 - The elucidation of the biosynthetic pathways of penicillin- and cephalosporin-producing microorganisms, and procuring of the genes involved in the biosynthesis of these β-lactam antibiotics have been facilitated by molecular biology techniques. Penicillins were firstly discovered in 1920s and reportedly synthesised by the fungus *Penicillium notatum.* Since then, more species of fungi and bacteria have been found to be capable of producing penicillins and some can even progress to transform these into cephalosporins, e.g., *Acremonium chrysogenum* and *Streptomyces clavuligerus*, respectively. In recent decades, the genetic engineering of procured β-lactam biosynthetic genes and the understanding of the mechanistic functions of the encoded key enzymes have stimulated advances in higher antibiotic productivity and in conjuring reactions to produce novel antibiotics. The introduction of additional copies of these genes or the whole penicillin gene cluster into *Penicillium chrysogenum* has also led to increased penicillin titres. These achievements attest to the possibility of improving strains through classical mutations and molecular modification of relevant biosynthetic genes. Due to its key role in the formation of the β-lactam ring, biochemical and structural studies of isopenicillin N synthase (IPNS) were also intensely pursued, facilitated by highly expressed recombinant proteins. After penicillin production, strains capable of expanding their prowess to cephalosporin production deviate from the penicillin producers by possessing another critical enzyme, deacetoxycephalosporin C synthase (DAOCS). This critical enzyme expands the thiazolidine ring of penicillins to the dihydrothiazine ring of cephalosporins. Remarkably, incorporation of the DAOCS coding gene from *S. clavuligerus* into *P. chrysogenum* made it possible to produce de novo, the cephalosporin precursor, 7-aminodeacetoxycephalosporanic acid (7-ADCA). Since the 'traditional' chemical synthesis of 7-ADCA is a difficult and expensive process, this genetic transformation of *P. chrysogenum* from a basic penicillin producer to an expanded cephalosporin producer has made it economical to derive 7-ADCA from its fermentation broths. As the 7-ADCA precursor is now readily available, its conversion to a myriad of semi-synthetic cephalosporins has become another valuable biosynthetic venture. Thus the interest in understanding the biosynthesis of cephalosporin has driven many molecular studies on IPNS and DAOCS to improve, modify and extend the availability of a plethora of β-lactam antibiotics. In this chapter, the authors discuss how molecular manipulation of the β-lactam biosynthetic genes has improved the production of certain penicillins and cephalosporins. Exploiting structural studies of recombinant IPNS and DAOCS in understanding the catalytic mechanism and substrate interaction of the β-lactam biosynthetic enzymes will also be reviewed.

Chapter 2 - The separation and recovery of pharmaceuticals such as penicillin G from fermentation media is very important. For this purpose, the research for an efficient, cost-effective and environmentally-friendly process continued over the last two decades. Various solvents including kerosene, butyl acetate, tributyl phosphate (TBP), octanol and their mixtures have shown to have the potential but they have the disadvantages of being toxic and expensive. Recently, non-traditional solvents like vegetable oils have been examined as they can overcome some of the above-mentioned disadvantages. In this chapter, results of research are presented for various organic solvents including non-traditional ones like vegetable oils for the separation of penicillin G using physical and reactive extractions. In the first step of the investigation, the effects of organic solvents on apparent distribution of penicillin G either from a single-component solution or from a synthetic media are determined. Then the effects of the addition of an ionic "carrier", such as Amberlite-LA- in the solvent on the enhancement of the distribution factor and on the degree of extraction are examined. The carrier-solvent systems have shown to have considerable improvement in the percentages of separation of penicillin G. As a demonstration of the improved process, the results are presented for Amberlite-LA2 in various organic solvents including sunflower oil phase. With the addition of a small amount (5%v/v) of the carrier in sunflower oil, the separation of penicillin G from the fermentation can be performed at the natural pH of fermentation (that is without adding any chemicals for pH adjustment). The percentage extraction of approx. 40-45% was achieved at a circulating flow rate of 13-16 L/h in a hollow-fibre membrane contactor. This result is similar to those reported by many investigators using toxic solvents (such as kerosene). The performance of the carrier-solvent system can be improved if pH and temperature are adjusted to lower values. This would require the media solution to be adjusted again for the fermentation of penicillin G as it requires a pH of 6.5. For recovery from the organic phase, aqueous solutions of sodium carbonate and sodium chloride have shown good results. An aqueous solution of sodium carbonate at pH 11 is able to recover significantly than sodium chloride solution and in a shorter process time of about 1.5 hours. From these studies it can be concluded that sunflower-based system is the preferred organic solvent for extraction because other solvents are very expensive (much more than that of sunflower oil) and do not have renewable sources. This new system offers many process advantages being less toxic, less corrosive, equally effective, environmentally-safe and operator-friendly.

Chapter 3 - Bacterial cell wall synthesis is a complex process that involves several key players. Among them, the Penicillin Binding Proteins (PBPs) and the ß-lactamases play a central role. The former are essential in cell wall synthesis. The latter appeared in response to decades of use of ß-lactam ring-based biomolecules acting as environmental selective pressures, and have become the most common efficient resistance systems developed by bacteria. The authors' laboratory pioneered the generation of catalytic antibodies using the anti-idiotypic approach. Using this approach, the authors engineered a catalytic antibody displaying a ß-lactamase-like activity: 9G4H9. This antibody was shown to be the internal image of class A ß-lactamase, the starting antigen of the idiotypic cascade. By using this catalytic antibody as the target, the authors selected a cyclic heptapeptide (Pep90) from a phage-displayed library, and studied its biochemical properties. Not only does Pep90 bind the catalytic antibody, but it is able to competitively inhibit its catalytic activity through key residues. Interestingly, Pep90 displayed striking inhibitory properties. Indeed, Pep90 was able to inhibit both TEM-1 ß-lactamase and some PBPs in a competitive manner. Few molecules are known to inhibiting both the antibiotic target (PBPs) and the resistance proteins (ß-lactamase). A molecule that displays both antibacterial and antibiotic properties, could potentially be used as a scaffold to generate peptidomimetic compounds. Indeed, the weakness of peptides as therapeutic agents is their tendency to undergo rapid proteolysis in serum. In order to overcome this limitation, the authors designed two peptidomimetics derived from Pep90 that contain unnatural amino_acids, and characterized their ability to inhibit both PBPs and ß-lactamase as peptidomimetic-like molecules. The authors' results suggest that Pep90 and its peptidomimetic counterparts may be good candidates for the development of new antibiotic molecules.

Chapter 4 - Penicillin derivatives are widely used for treatment of bacterial infection. A high-dose penicillin is recommended as a first choice particularly for infective endocarditis in the guideline from the American Heart Association. The authors found two patients with infective endocarditis who had serum albumin shifting to a cathode side on the electrophoresis. The both patients had been treated with a high-dose penicillin G. The serum albumin concentrations of the patients were 1.7 g/dL and 1.8 g/dL when measured by a modified bromocresol purple (BCP) method while they were 3.4 and 3.3, respectively, by a bromocresol green (BCG) method. To confirm these low values obtained by the modified BCP method are due to the high-dose penicillin G therapy, the authors added penicillin G in normal sera and measured albumin concentration by the BCP and BCG methods. Exogenously added penicillin G inhibited the BCP reaction in dose-dependent and incubation time-dependent manners but not the BCG reaction. Here, the authors generally describe effects of penicillin derivatives on serum albumin measurement and other clinical tests.

Chapter 5 - Effects of ozonation and Fenton's treatment on the biodegradability of pharmaceutical wastewater originating from the *Sultamycillin Tosylate Diydrate* penicillin formulation was investigated. Process variables were ozone feed rates (127 – 2750 mg $l^{-1}h^{-1}$), ozonation pH as well as the type of Fenton's reagents (Fenton; Fe^{2+}/H_2O_2 and Fenton-like; Fe^{3+}/H_2O_2) applied at a fixed pH (= 3.0) and different Fe (0.5, 1 mM Fe^{2+} and Fe^{3+}) and H_2O_2 concentrations (10, 20 mM). The efficiency of the selected pre-treatment systems was evaluated in terms of changes in COD (Chemical Oxygen Demand), TOC (Total Organic Carbon) and BOD_5 (5-Day Biochemical Oxygen Demand). The effect of chemical oxidative pre-treatment on activated sludge treatment performance was assessed in terms of microbial oxygen uptake rate (OUR) measurements conducted with unacclimated sewage sludge. Results have demonstrated that although similar overall COD (34 - 42 %) and TOC (24 - 36 %) removal efficiencies have been obtained by the ozonation process and the Fenton and Fenton – like reagents, only ozonation and Fenton-like processes were capable of completely eliminating the inhibitory effect of the penicillin formulation effluent on the microbial consortium in sewage sludge. Moreover, the Fenton-like and ozonation processes appeared to slightly enhance COD abatement rates by 4 % and 8 %, respectively, as could also be demonstrated in separate biodegradation experiments that were run with the raw, pre-treated formulation effluent in the same biosludge.

Chapter 6 - Quantum chemical calculation of a molecule of benzilpenicillin is executed by method AB INITIO in base 6-311G** with optimization of geometry on all parameters for the first time. The optimized geometrical and electronic structure of this compound is received. Acid force of benzilpenicillin is theoretically appreciated. It is established, than it to relate to a class of weak H-acids (pKa=+12, where pKa-universal index of acidity).

In: Penicillin
ISBN: 978-1-62100-233-8
Editors: G. Barresi and P. Komulainen

Chapter 1

Molecular Modification of Biosynthetic Genes for Improved Production of Penicillins and Cephalosporins

K. S. Goo, D. S. H. Tan and T. S. Sim[*]
Department of Microbiology, Yong Loo Lin School of Medicine,
National University of Singapore, Singapore

Abstract

The elucidation of the biosynthetic pathways of penicillin- and cephalosporin-producing microorganisms, and procuring of the genes involved in the biosynthesis of these β-lactam antibiotics have been facilitated by molecular biology techniques. Penicillins were firstly discovered in 1920s and reportedly synthesised by the fungus *Penicillium notatum*. Since then, more species of fungi and bacteria have been found to be capable of producing penicillins and some can even progress to transform these into cephalosporins, e.g., *Acremonium chrysogenum* and *Streptomyces clavuligerus*, respectively. In recent decades, the genetic engineering of procured β-lactam biosynthetic genes and the understanding of the mechanistic functions of the encoded key enzymes have stimulated advances in higher antibiotic productivity and in

[*] Corresponding author. Fax: (65)-6776 6872; Tel: (65)-6516 3280; E-mail: micsimts@nus.edu.sg.

conjuring reactions to produce novel antibiotics. The introduction of additional copies of these genes or the whole penicillin gene cluster into *Penicillium chrysogenum* has also led to increased penicillin titres. These achievements attest to the possibility of improving strains through classical mutations and molecular modification of relevant biosynthetic genes. Due to its key role in the formation of the β-lactam ring, biochemical and structural studies of isopenicillin N synthase (IPNS) were also intensely pursued, facilitated by highly expressed recombinant proteins. After penicillin production, strains capable of expanding their prowess to cephalosporin production deviate from the penicillin producers by possessing another critical enzyme, deacetoxycephalosporin C synthase (DAOCS). This critical enzyme expands the thiazolidine ring of penicillins to the dihydrothiazine ring of cephalosporins. Remarkably, incorporation of the DAOCS coding gene from *S. clavuligerus* into *P. chrysogenum* made it possible to produce de novo, the cephalosporin precursor, 7-aminodeacetoxycephalosporanic acid (7-ADCA). Since the 'traditional' chemical synthesis of 7-ADCA is a difficult and expensive process, this genetic transformation of *P. chrysogenum* from a basic penicillin producer to an expanded cephalosporin producer has made it economical to derive 7-ADCA from its fermentation broths. As the 7-ADCA precursor is now readily available, its conversion to a myriad of semi-synthetic cephalosporins has become another valuable biosynthetic venture. Thus the interest in understanding the biosynthesis of cephalosporin has driven many molecular studies on IPNS and DAOCS to improve, modify and extend the availability of a plethora of β-lactam antibiotics. In this chapter, we discuss how molecular manipulation of the β-lactam biosynthetic genes has improved the production of certain penicillins and cephalosporins. Exploiting structural studies of recombinant IPNS and DAOCS in understanding the catalytic mechanism and substrate interaction of the β-lactam biosynthetic enzymes will also be reviewed.

Keywords: β-lactam, penicillin, cephalosporin, isopenicillin N synthase, deacetoxycephalosporin C synthase

Introduction

One of the most significant advances made in medical history is the discovery of penicillin. It paved the way for the chemotherapeutic treatment of infectious diseases, many of which were left untreated prior to the discovery. Penicillin was discovered through the keen observation of Alexander Fleming when he noticed that a clinical isolate of *Staphylococcus aureus* was inhibited by a mould contaminant on an agar plate. The contaminant was identified as *Penicillium notatum* and the antibacterial compound was named penicillin, after the producing strain (Fleming, 2001).

The early success of penicillin in the treatment of infectious diseases spurred researchers, especially those in the pharmaceutical companies to develop screening programmes directed towards similar antibiotics. It became apparent in subsequent findings that the key efficacy of these antibiotics was dependent on the β-lactam ring structure. Rational screening programmes were developed using mutants of bacteria, e.g., *Escherichia coli* and *Pseudomonas* spp., which were highly susceptible to β-lactam antibiotics (Aoki et al., 1976; Kitano et al., 1975). Such intensive efforts led to the discovery of other strains of fungi and a number of bacteria, especially those from the actinomycete family, that are capable of producing a plethora of β-lactam antibiotics (Demain, 1983). The sulphur containing β-lactam antibiotics, comprising mainly penicillins and cephalosporins, are still the most commonly prescribed antibiotics due to their lack of toxicity coupled with potent antibacterial activities. They eradicate bacteria by specifically targeting the penicillin-binding proteins (PBPs) resulting in the inhibition of bacterial cell wall synthesis. They are relatively non-toxic because mammalian cells do not possess PBPs (Demain and Elander, 1999). Hence, the β-lactam compounds account for over 65% of the total antibiotic market for treatment of systemic diseases (Schmidt, 2010).

Penicillin and cephalosporin antibiotics, also known as penams and cephems, respectively, are structurally related. Both contain a β-lactam ring fused to a second sulphur-containing ring. The presence of the sulphur atom in the second ring distinguishes the penicillins and cephalosporins from other classes of β-lactam antibiotics like the clavams, carbapenems and monobactams (Figure 1). While the β-lactam ring is fused to a five-membered thiazolidine ring in penicillins, it is fused to a six-membered dihydrothiazine ring in cephalosporins. Currently, there are at least three different classes of cephem antibiotics. Cephalosporin C is the first cephalosporin discovered and so named after the earlier genus name of the producer *Acremonium chrysogenum*, i.e., *Cephalosporium acremonium* (Campos Muñiz et al., 2007). Cephamycin C deviates from cephalosporin C by possessing a methoxy group at C7α and a carbamoyl group at C3 instead of an acetyl group (Figure 2). So far, it is mainly produced by actinomycetes viz *Nocardia lactamdurans* and *Streptomyces* species (Nagarajan et al., 1971; Stapley et al., 1972). The third group of cephalosporin is cephabacin. Similar to cephamycin C, cephabacin differs from cephalosporin C at its C7α and C3 positions. It possesses a

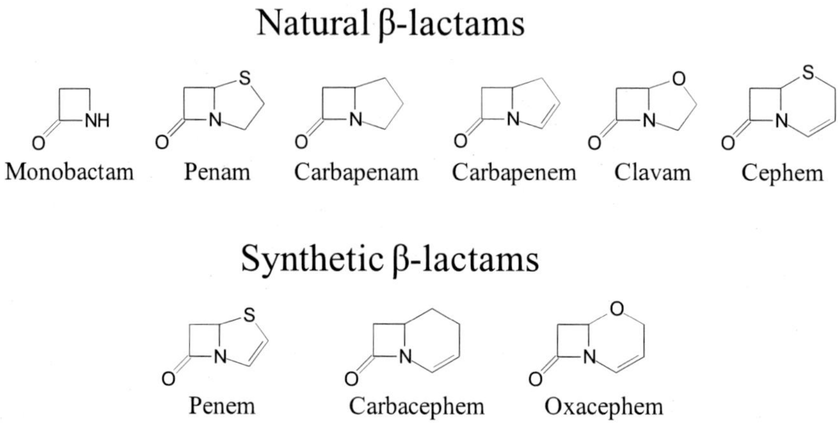

Figure 1. Basic structures of various classes of β-lactam antibiotics.

formylamino group at C7α and an oligopeptide moiety at its C3 position. Presently, cephabacins are only known to be produced by Gram-negative bacteria *Lysobacter lactamgenus* (Ono et al., 1984), *Xanthomonas lactamgena* (Ono et al., 1984) and *Flavobacterium* sp. (Palissa et al., 1989). Both cephamycin C and cephabacin have been demonstrated to be more resistant to the degradation of β-lactamases than cephalosporin C (Nozaki et al., 1985; Nozaki et al., 1984).

The overwhelming success of the β-lactam antibiotics had impelled both industrial and academic laboratories to focus intense efforts towards understanding the biosynthesis and production of β-lactam antibiotics. Improvement of antibiotic yield was the main target of research in the early decades, and approaches such as natural selection and extensive mutagenesis programmes coupled with biochemical screening were used (Demain, 1973). Efforts were also directed at obtaining metabolically blocked mutants, which provided some of the necessary background for mapping out the biosynthesis of the β-lactam antibiotics. The biosynthetic pathways of penicillin G in *Penicillium chrysogenum*, cephalosporin C in *A. chrysogenum*, cephamycin C in *Streptomyces clavuligerus* and cephabacins in *L. lactamgenus* were subsequently elucidated (Figure 2).

Further advancement in the use of penicillins was achieved by the modification of the side chain of penicillin leading to novel penicillin compounds with improved efficacy. This was possible through semi-synthetic methods in which various side chains were added to the nucleus of the penicillin molecule, 6-aminopenicillanic acid (6-APA). As reported, the production of 6-APA can be achieved by fermentation in the absence of side-chain precursors, by hydrolysing benzylpenicillin using bacterial amidases and by chemical synthesis (Wolfe et al., 1984). As a result of the discovery of 6-APA, many semi-synthetic β-lactam antibiotics with increased efficacy such as ampicillin, amoxicillin and cloxacillin were generated (Nayler, 1991).

Following the elucidation of the structure of cephalosporin C, an entirely new group of clinically significant β-lactam antibiotics was subsequently produced, i.e., the cephem antibiotics. The success of 6-APA led to the production of the cephalosporin nucleus, 7-aminocephalosporanic acid (7-ACA), and this was used for the semi-synthesis of many cephalosporin antibiotics, which today plays a crucial role in the treatment of infectious diseases (Hamilton-Miller, 2008). These cephalosporin antibiotics were demonstrated to have a broader antimicrobial spectrum than the penicillins; more importantly, they possessed greater resistance to the penicillinases produced by many penicillin-resistant pathogens.

In an effort to develop strains of bacteria and fungi with improved production of β-lactam compounds as well as novel β-lactam compounds, recombinant DNA technology was used for more precise genetic manipulation of the β-lactam producing microorganisms (Wolfe et al., 1984). The application of molecular biology techniques to the study of the β-lactam antibiotics thus led to the cloning of many of their biosynthetic genes. This has contributed to an improved understanding of both the β-lactam biosynthetic genes and their encoded enzymes. At the same time, it has opened up the possibility of improving key enzymes for improved yield or production of novel compounds.

The β-lactam antibiotics continue to be popularly used due not only to their therapeutic effectiveness and reduced toxicity but also to their abundant availability at reduced pricing. This is because of the improved developments in fermentation process, the genetic improvement of producing strains, and downstream processing (Schmidt, 2010). Moreover, the development of semi-synthetic penicillins and cephalosporins expanded the range and effectiveness of these β-lactam antibiotics for chemotherapeutic use (Hamilton-Miller, 2008).

However, the chemotherapeutic application of the β-lactam antibiotics has been threatened by the increasing occurrence of bacteria resistant to the β-lactam antibiotics. According to the World Health Organisation, acute infectious diseases cause 25% of deaths worldwide, resulting in 13–17 million fatalities per year (Gannon, 2000). In the face of widespread antibiotic resistance, threat of emerging and re-emerging pathogens, and the consequential high social and economic impact of infectious diseases, there is a continuing challenge to pursue better strategies to counteract these antagonistic microorganisms.

BIOSYNTHESIS OF β-LACTAM ANTIBIOTICS

The primary interest in understanding β-lactam biosynthesis originated from the structural similarity between penicillin and cephalosporin. Resolution of the chemical structure of cephalosporin C bearing a β-lactam ring fused with a dihydrothiazine ring further affirmed the possible genetic linkage between their biosyntheses (Abraham and Newton, 1961). Studies on the biosynthetic pathways of penicillin and cephalosporin producers mainly concentrated on *P. chrysogenum*, *A. chrysogenum*, *S. clavuligerus*, *N. lactamdurans* and *L. lactamgenus*. Despite 30 years of investigation, the authors' understanding of the biosynthetic pathway of β-lactams is not complete because of branching into varied pathways leading to different end-products in different microorganisms, especially the cephamycin and cephabacin producers (Liras, 1999; Liras and Demain, 2009; Martin and Gutierrez, 1995; Martin et al., 2010). Nevertheless, the major pathway leading to the formation of the penicillin and cephalosporin core structures have been deciphered, allowing genetic manipulation of industrial strains for high yield production (Figure 2).

Biosynthesis of Penicillin

Biosynthesis of penicillin begins with the fusion of three amino acids, i.e., L-aminoadipic acid, L-cysteine and L-valine, into a tripeptide known as δ-(L-α-aminoadipyl)-L-cysteinyl-D-valine (ACV), by ACV synthetase (ACVS) encoded by *pcb*AB (Coque et al., 1991; Diez et al., 1990; Gutierrez et al., 1991; Loder and Abraham, 1971; Tobin et al., 1991). ACV is cyclised to isopenicillin N by isopenicillin N synthase (IPNS), which is encoded by *pcb*C gene (Castro et al., 1988; Diez et al., 1989). This is the formation of a first bioactive compound in the biosynthetic pathway. IPNS is a non-haem ferrous iron-dependent oxygenase, which bears the conserved iron-binding motif HXD/EX$_n$H. It also possesses the RXS motif, which is critical for the coordination of the co-substrate 2-oxoglutarate in the enzymatic reaction of 2-oxoglutarate dependent dioxygenases. Interestingly, IPNS does not require 2-oxoglutarate for its reactivity. Instead, the RXS motif is important for binding its valine carboxylate of ACV substrate (Roach et al., 1995; Roach et al., 1997).

In penicillin-producing organisms, e.g., *P. chrysogenum*, the end-product of its β-lactam biosynthetic pathway can be altered by feeding phenylacetic acid or phenoxyacetic acid to the fermentation culture to derive penicillin G or penicillin V, respectively, which are the earliest clinically useful penicillins. This is achieved by acyl-coenzyme A:IPN acyltransferase, which exchanges the aminoadipyl side chain of isopenicillin N to the respective coenzyme A-activated feeding substrate (Barredo et al., 1989) (Figure 2). This enzyme not only possesses IPN acyltransferase activity, it also bears amidohydrolase and 6-APA acyltransferase activities, and accepts a wide range of acyl-CoA derivatives and various non-CoA thioesters, highlighting a potential application in the synthesis of semi-synthetics (Alvarez et al., 1993; Whiteman et al., 1990). Although this enzyme is encoded by a single *pen*DE gene in both fungi, *P. chrysogenum* IPN acyltransferase is auto-processed from a 40 kDa inactive proenzyme into a heterodimer consisting of α (29 kDa) and β (11 kDa) subunits, which differs from *Aspergillus nidulans* IPN acyltransferase that is active in its unprocessed form. This difference in functionality has been attributed to the protein sequence disparity in the proenzyme, which determines the autocatalytic activity (Fernandez et al., 2003).

Biosynthesis of Cephalosporin C

Studies on the penicillin- and cephalosporin-biosynthetic pathways in producing organisms have revealed a common prelude between them (Brakhage et al., 2005; Liras, 1999; Martin, 1998; Queener, 1990) (Figure 2). In cephalosporin C-producing *A. chrysogenum*, isopenicillin N is converted into penicillin N by a two-component isopenicillin N epimerase system. This two-component protein system is encoded by two genes, namely *cef*D1 and *cef*D2, which correspond to isopenicillinyl-CoA ligase and isopenicillinyl-CoA epimerase, respectively (Martin et al., 2004; Ullan et al., 2002). Penicillin

N is then ring-expanded to deacetoxycephalosporin C (DAOC) by a bifunctional deacetoxy/deacetylcephalosporin C synthase (DAOC/DACS) enzyme, where the five-membered thiazolidine ring in penicillin N is converted to the six-membered dihydrothiazine ring in DAOC (Dotzlaf and Yeh, 1989). DAOC is subsequently hydroxylated at C3 position by the same enzyme to yield deacetylcephalosporin C (DAC). Then an acetyl group is added to the C3 hydroxyl group of DAC by DAC acetyltransferase to finally produce cephalosporin C (Gutierrez et al., 1992).

Biosynthesis of Cephamycin C

In *S. clavuligerus* and *N. lactamdurans*, isopenicillin N is converted to penicillin N by a monomeric pyridoxal phosphate-dependent isopenicillin N epimerase, which is different from the *A. chrysogenum* epimerase system (Martin et al., 2004; Usui and Yu, 1989). Penicillin N is ring-expanded to DAOC followed by hydroxylation to DAC using two separate monofunctional deacetoxycephalosporin C synthase (DAOCS) and deacetylcephalosporin C synthase (DACS) enzymes (Coque et al., 1996; Cortes et al., 1987; Jensen et al., 1985). These two steps are similar to that of the cephalosporin C biosynthetic pathway. Thereafter, a carbamoyl group is added to the C3 hydroxyl of DAC by DAC-carbamoyl transferase (Coque et al., 1995b), followed by methoxylation at C7α by O-carbamoyl-DAC-7-methoxyl transferase (Coque et al., 1995a). Studies have shown that the hydroxyl and methyl groups are derived from molecular oxygen and methionine, respectively (O'Sullivan et al., 1979; Whitney et al., 1972). This last step of methoxylation is carried out by two enzymes CmcI and CmcJ separately encoded by the *cmc*I and *cmc*J genes. These two enzymes function as a duo-protein component in hydroxylating the C7α position of the cephem nucleus followed by methylation, although in *N. lactamdurans*, the CmcI enzyme alone has weak hydroxylation activity in the presence of 2-oxoglutarate (Coque et al., 1995a; Enguita et al., 1996). Interestingly, *N. lactamdurans* CmcI was also found to exhibit hydroxylase activity in the presence of NADH although no methyltransferase activity was detected for the CmcI-CmcJ complex (Coque et al., 1995a). Fluorescence spectroscopic analysis of CmcI and CmcJ interaction illustrated conformational changes in CmcI with increasing concentrations of CmcJ, and the binding of substrates S-adenosylmethionine and cephalosporin C was higher in the CmcI-CmcJ complex compared to the individual proteins (Enguita et al., 1996).

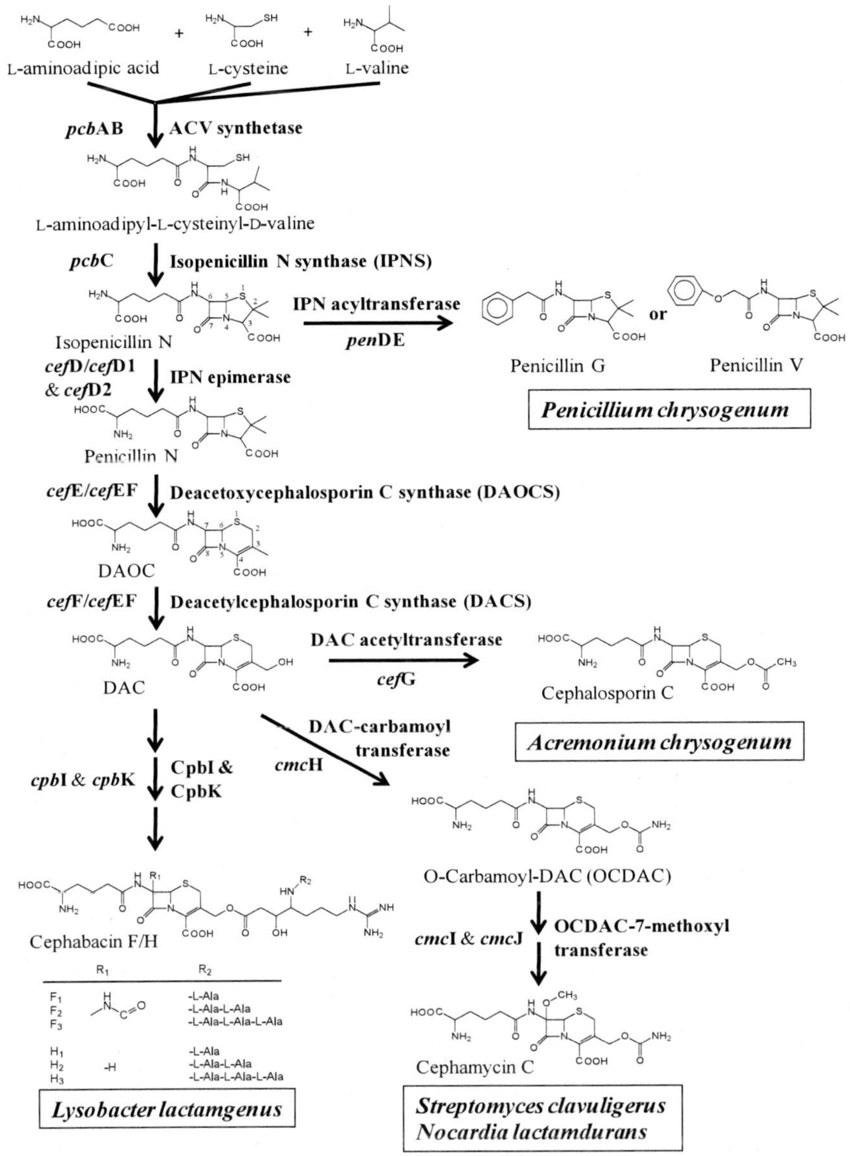

Figure 2. Biosynthesis of cephalosporin C, cephamycin C and cephabacin in respective microorganisms. In *A. chrysogenum*, the epimerisation of isopenicillin N to penicillin N involves two different enzymes encoded by *cef*D1 and *cef*D2 genes, and the ring-expansion followed by hydroxylation of penicillin N to DAC is catalysed by a bifunctional enzyme DAOC/DACS encoded by *cef*EF gene.

Biosynthesis of Cephabacin

Similar to the biosynthesis of cephamycin C, the biosynthesis of cephabacin differs from that of cephalosporin C after DAC is formed although it is still not clear if the formylamino or the oligopeptide moiety was added first (Figure 2). Recent efforts in deciphering the functionalities of enzymes involved in the synthesis of the oligopeptide moiety have shed light on their identities. The *cpb*I gene encodes the non-ribosomal peptide synthetase-polyketide synthase-hybrid complex (NRPS-PKS), in which the NRPS component is composed of three modules. Each of these modules has three domains responsible for catalysing condensation, adenylation and thiolation reactions. The *cpb*K gene encodes only a single NRPS module (Demirev et al., 2006). These two genes encode enzymes necessary for the formation of the oligopeptide moiety. However, the enzymes responsible for the addition of the formylamino and oligopeptide moieties to the C7α and C3 position of cephabacin F, respectively, has yet to be elucidated.

Development of Semi-synthetic Penicillins

Natural penicillins are structurally diverse and their biosyntheses are dependent on the types of side chain precursors available. The first clinically useful penicillin G was produced by supplying phenylacetic acid in the fermentation culture of *P. chrysogenum*. However, it was only used as a parenteral chemotherapeutic because of its instability in the acidic gastric conditions. Penicillin V was found to be more acid-stable and hence, could be administered orally. This attribute was mainly due to the C6 phenoxyacetyl side chain in penicillin V compared to penicillin G. However, both penicillins have narrow antimicrobial spectra and were only effective in treating diseases caused by Gram positive bacteria (Wright, 1999). Following the use of penicillin G and penicillin V, the emergence of penicillin-resistant staphylococci became a serious clinical problem (Chain, 1972). This resistance factor was found to be an inactivator of penicillin and was termed as penicillinase (Kirby, 1944). Thereafter, many studies were pursued with the intention of improving the antimicrobial spectrum of penicillin to include Gram negative bacteria as well as penicillin-resistant staphylococci.

Recognizing the importance of C6 side chain of penicillin for its antibacterial property, successes in the development of more efficacious penicillins were brought about by the discovery of 6-APA production in the fermentation culture of *P. chrysogenum* (Rolinson and Geddes, 2007) (Figure 3). By chemically varying the structure of its C6 side chain, a structurally diverse group of semi-synthetic penicillins could be synthesised with different antimicrobial spectra. The semi-synthetic penicillins are mainly classified by structural similarity, which also reflects their antimicrobial spectra.

Methicillin is the first semi-synthetic penicillin that has an ortho-dimethoxyphenyl group as a side chain and was resistant to the action of staphylococcal penicillinase (Rolinson, 1998; Wright, 1999). It was clinically useful in inhibiting penicillin-resistant *S. aureus* in the early 1960s but its inhibitory activity was not as high as penicillin G against similarly tested susceptible pathogens. It was eventually replaced by other penicillinase-resistant penicillins e.g. oxacillin and cloxacillin, that have higher bioactivity against staphylococci (Wright, 1999). However, strains of *S. aureus* that were resistant to methicillin and other penicillinase-resistant penicillins eventually emerged and till now, still pose a problem to the healthcare industries (Appelbaum, 2006). They are known as methicillin-resistant *S. aureus* (MRSA) and the resistant mechanism they possessed was due to the acquisition of a low affinity PBP to β-lactam antibiotics known as PBP2a encoded by *mec*A, instead of evolved β-lactamases (Llarrull et al., 2009; Matsuhashi et al., 1986; Ubukata et al., 1989; Ubukata et al., 1985).

Ampicillin is the first semi-synthetic penicillin with a broader antimicrobial spectrum than penicillin G, in that it inhibits various Gram negative bacteria (Rolinson, 1998). It has a phenylglycyl side chain, which structurally differs from penicillin G by an amino group in the α-carbon of the side chain. However, it has a poor absorption property in the alimentary canal. Subsequently, the addition of a hydroxyl group in the *para* position of the aromatic ring of ampicillin converted it to a more efficacious amoxicillin. Amoxicillin had better absorption from the gut and resulted in higher levels in the blood than equivalent doses of ampicillin could (Wright, 1999). However, resistance to amoxicillin due to β-lactamases also became wide-spread in the 1970s. Fortunately, a β-lactamase inhibitor, which has a highly similar chemical structure to penicillin, was discovered. It was mainly produced by *S. clavuligerus* and was named clavulanic acid. This discovery led to the first combinatorial administration of amoxicillin and clavulanic acid, a β-lactamase inhibitor, to overcome the proliferation of β-lactamase-producing pathogens (Miller et al., 2001; Wright, 1999).

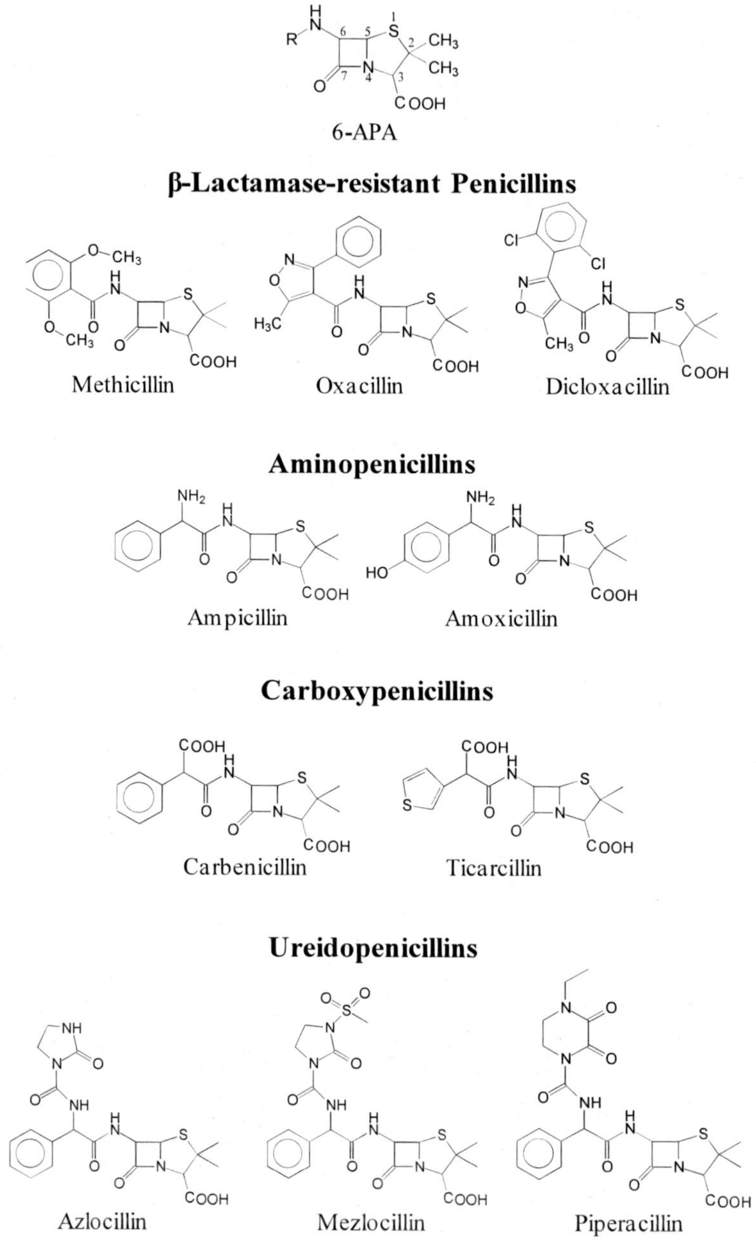

Figure 3. Chemical structures of penicillin nucleus 6-aminopenicillanic acid (6-APA) and major groups of semi-synthetic penicillins.

Although both ampicillin and amoxicillin were considered to have broad antimicrobial spectrum, they do not have good inhibitory activity against *P. aeruginosa*, which is one of the leading causes of nocosomial infections, especially in immunocompromised and cystic fibrosis patients (Mesaros et al., 2007; Strateva and Yordanov, 2009). By replacing the amino group with a carboxyl group in the side chain of ampicillin, improved activity against *P. aeruginosa* was observed and this penicillin was named carbenicillin. Then by replacing the phenyl ring with a thiophenyl ring in carbenicillin, ticarcillin was found to possess better anti-pseudomonal activity than carbenicillin (Rolinson, 1998). By virtue of the carboxyl group in the side chain of both penicillins, they are classified as carboxypenicillins.

Subsequently, the addition of urea moieties to the α-amino group in the side chain of ampicillin yielded ureidopenicillins (azlocillin, piperacillin and mezlocillin), which have even higher activities against *P. aeruginosa*, while retaining its inhibitory activity against streptococci, enterococci and non-β-lactamase-producing staphylococci (Bush and Johnson, 2000; Tan and File, 1995; Wright, 1999). However, the development of semi-synthetic cephalosporins eventually superseded that of the penicillins probably due to the amenability of the cephalosporin nucleus for multiple chemical modifications.

Genetic Manipulation of β-lactam Producers to Improve Penicillin Production Yield

Penicillin Production Improved by Classical Strain-Improvement

As Fleming's original *P. notatum* strain (NRRL 1249B21) only produced low amounts of penicillin (about 1.2 μg/mL), there was great interest in isolating new strains with improved penicillin production. This led to the discovery of *P. chrysogenum* NRRL 1951 from a mouldy cantaloupe in Illinois, USA in 1943 (Raper et al., 1944). Submerged cultures of this strain could produce about 100-fold more than that of *P. notatum*. Subsequently, through classical mutagenesis of *P. chrysogenum* NRRL 1951 at the University of Wisconsin, an improved producer strain Wisconsin 54-1255 that was capable of producing 550 μg/mL penicillin was identified (Elander,

1983). Strain improvement programmes undertaken by various pharmaceutical companies followed and led to the development of higher-producing *P. chrysogenum* strains such as the P2 strain of Panlabs (Taiwan) (Lein, 1986), the DS04825 strain obtained at DSM (The Netherlands) (Gouka et al., 1991), the AS-P-78 and the E1 strains, which were obtained at Antibioticos S.A. (Spain) (Fierro et al., 1995) and the BW1952 strain at SmithKline Beecham (UK) (Newbert et al., 1997). These industrial strains are capable of producing more than 50, 000 μg/mL of penicillin.

As empirical mutagenic hits are random and may affect genes other than those involved in antibiotic biosynthesis, higher penicillin-producing strains obtained can be rather genetically unstable. Thus, routine re-isolation is required for selecting the best performing strains industrially (Barber et al., 2004). Despite the process being laborious and random, it was the common approach applied in *Penicillium* industrial strain improvement programmes until the advent of recombinant DNA techniques, which brought about the possibility of using different strategies to improve penicillin production via directed gene manipulations.

Application of Molecular Biology Techniques for Improved Penicillin Production

Molecular biology represents an advanced tool in the genetic improvement of industrial producer strains. These techniques allow specific modifications in the genome of the microorganisms to be carried out. With the genes responsible for the biosynthesis of antibiotics cloned, improvement to the producing strains was based on the idea that if production depended upon the expression of these genes, any change capable of positively affecting that expression would be translated into increased production. Initial application of recombinant DNA techniques to increase antibiotic production thus consisted of increasing the dose of antibiotic biosynthetic genes, thereby increasing the expression of the corresponding biosynthetic enzymes (Table 1).

After the *pcbC* gene encoding for IPNS was isolated from *P. chrysogenum* and other penicillin producers, the fungal *pcbC* gene was reintroduced into an industrial production strain of *P. chrysogenum* (Skatrud et al., 1987). However, no increase in penicillin production could be demonstrated in the transformants analysed. With the *pen*DE gene from *P. chrysogenum* subsequently isolated (Barredo et al., 1989), a fragment containing both the *pcbC* and *pen*DE genes was used to transform the *P. chrysogenum* Wisconsin 54-1255 strain (Veenstra et al., 1991). This resulted in a significant 40% increase in penicillin G production. It seemed that the amplification of both genes and their cooperative expression are necessary to obtain an improvement in penicillin production. This was later confirmed when the complete cluster for penicillin biosynthesis (*pcb*AB, *pcb*C and *pen*DE genes) was introduced into a low-producing laboratory strain of *P. chrysogenum* (Theilgaard et al., 2001). Following the introduction of different combination of the three structural genes, Theilgaard and co-workers (2001) showed that a fine balance exists between ACVS and IPNS, which appears to control the metabolic flux through the pathway.

Recently, Nijland and co-workers (2010) demonstrated that there is a limit to improving production via amplification of genes in the penicillin biosynthetic cluster. They analysed the gene dosage effect of three key enzymes, ACVS, IPNS and IPN acyltransferase, using the high-producing strain DS17690 devoid of the penicillin cluster as the basic host. At higher copy numbers of the cluster, although isopenicillin N levels were increased, the penicillin yield did not increase accordingly indicating that the acyltransferase reaction was the rate-limiting step (Nijland et al., 2010).

In *A. nidulans,* the expression levels of the three structural genes (*pcb*AB, *pcb*C, and *pen*DE) were amplified by promoter replacement with the strong inducible ethanol dehydrogenase promoter (*alc*A) (Kennedy and Turner, 1996; Penalva et al., 1998). While the overexpression of IPNS alone resulted in only a marginal increase in penicillin production, the overexpression of ACVS led to a 30-fold increase in penicillin G yield (Kennedy and Turner, 1996). When both the ACVS and IPNS genes were overexpressed in *A. nidulans*, a marked increase in penicillin production was achieved over that which overproduced ACVS alone (Penalva et al., 1998). It appears again that the simultaneous overexpression of the biosynthetic genes is required to improve the antibiotic titre, presumably by moderating the metabolic flux.

As the formation of penicillin begins with three amino acids, it was suggested that the increase of these amino acid pools might lead to increase antibiotic production. In the filamentous fungi, the lysine pathway is involved in producing aminoadipic acid (AAA) as an intermediate. In *P. chrysogenum,* the deletion of the *lys*2 gene encoding aminoadipate reductase was shown to lead to an increase of the precursor AAA and subsequently a doubling of the penicillin G production (Casqueiro et al., 1999).

Table 1. Genetic manipulation of β-lactam producers in forging improved production of penicillin

Strain	Method	Improvement	Remarks	References
P. chrysogenum	Introduction of *pcbC* gene encoding IPNS	No change	No increase in penicillin G production	Skatrud et al., 1987
P. chrysogenum	Introduction of a fragment containing both the *pcbC* and *pen*DE genes	Higher penicillin G production	Resulted in a significant 40% increase in penicillin G production	Veenstra et al., 1991
P. chrysogenum	Introduction of PCL gene encoding phenylacetyl CoA ligase	Increase in precursor phenylacetyl CoA	Penicillin G production doubled	Minambres et al., 1996
A. nidulans	Use of strong inducible ethanol dehydrogenase promoter to express the three structural genes (*pcb*AB, *pcb*C, and *pen*DE)	Higher penicillin G production	30-fold increase in penicillin yield over strains overproducing *acv*A alone	Penalva et al., 1998
P. chrysogenum	Deletion of *lys*2 gene encoding aminoadipate reductase	Increase in precursor AAA produciton	Increased AAA level led to a double increase in penicillin G production	Casqueiro et al., 1999
P. chrysogenum	Introduction of *pcb*AB, *pcb*C, and *pen*DE genes	Higher penicillin G production	Transformants with the whole gene cluster amplified showed the largest increase in specific penicillin productivity -124% and 176%, respectively	Theilgaard et al., 2001

In the formation of penicillin G, the precursor phenylacetyl CoA is required. Another molecular approach to augment the penicillin yield in *P. chrysogenum* was to increase the amount of phenylacetyl CoA precursors by overexpressing the phenylacetyl CoA ligase (PCL). The PCL from

Pseudomonas putida was found to be able to activate phenylacetic acid to phenylacetyl CoA at a rate of more than 500-fold higher that other acyl-CoA-activating enzymes (Luengo, 1995). When additional copies of the PCL gene from *P. putida* were cloned into *P. chrysogenum* Wisconsin 54-1255, penicillin G production doubled (Minambres et al., 1996).

DEVELOPMENT OF SEMI-SYNTHETIC CEPHALOSPORINS

Although cephalosporin C is the first cephalosporin discovered, its narrow antimicrobial spectrum limited its clinical application. Nevertheless, the discovery of cephalosporin C led to the development of a plethora of cephalosporin derivatives. It was recognised that, similar to the development of semi-synthetic penicillins, the availability of the cephalosporin core structure was important in modifying the side chain of cephalosporin to improve its spectrum and efficacy. The production of the cephalosporin core structure, i.e., 7-aminocephalosporanic acid (7-ACA), was made possible by deacylation of the aminoadipyl side chain of cephalosporin C. Hence, 7-ACA was used as a precursor for the synthesis of many earlier semi-synthetic cephalosporins (Hamilton-Miller, 2008).

Other than having a similar chemical structure and mode of action as penicillin, cephalosporins were found to be naturally resistant to penicillinase. These two attributes make it a more attractive lead compound in the development of new β-lactam antibiotics (Hansch et al., 1990). Furthermore, cephalosporin has two alkyl sites amenable to chemical modification compared to penicillin, which only has one (Figure 4). This allows a larger variety of cephalosporin derivatives to be synthesised and assessed for better antimicrobial properties for clinical applications. Hence, a plethora of β-lactam derivatives were developed and marketed, with an estimation of at least 50 different cephalosporins available in the market in 2003 (Elander, 2003; Hamilton-Miller, 2000, 2008).

Cephalexin is one of the earlier semi-synthetic cephalosporin available and has the same phenylglycyl side chain as ampicillin. Its antimicrobial spectrum encompasses that of ampicillin except that its resistance to penicillinase bestowed itself a wider spectrum than ampicillin (Hansch et al., 1990). Similarly, cefadroxil, which is a *p*-hydroxyl derivative of cephalexin, also possesses similar pharmacokinetic properties but wider spectrum than its penicillin counterpart, amoxicillin (Hansch et al., 1990). It became clear that in order to further improve the efficacy of cephalosporin, the structure of the side chains of cephalosporin has to be different from that of the semi-synthetic penicillins. Thereafter, the pool of structurally diverse semi-synthetic cephalosporins with different antimicrobial spectra and susceptibilities to the β-lactamases increased tremendously.

The semi-synthetic cephalosporins are generally grouped into 'generations' based on their antimicrobial spectra and stability to β-lactamases, which is considered more useful for administrative reference. The first generation cephalosporins are made up of early cephalosporins that have similar antimicrobial spectra as that of cephalexin. They were considered broad spectrum antibiotics with the advantage of being more stable to penicillinase. The second and third generations have extended spectrum of the first generation cephalosporins, including many Gram negative bacteria, and are more resistant to the evolving β-lactamases. However, these improved cephalosporins suffered lower inhibitory activity against Gram positive bacteria. In contrast to the first three generations, the fourth-generation cephalosporins, e.g., cefepime, have a broad antimicrobial spectrum including many Gram positive and Gram negative bacteria, as well as *P. aeruginosa*. They are also generally more resistant to the evolved β-lactamases active against earlier generation cephalosporins (Marshall and Blair, 1999). However, the issue with MRSA remained problematic until recently when new two cephalosporins, namely ceftobiprole (Deresinski, 2008; Llarrull et al., 2009) and ceftaroline (Llarrull et al., 2009; Villegas-Estrada et al., 2008), were developed. Both cephalosporins were tested and shown to be efficacious against MRSA due to higher binding affinity to the resistant PBP2A. Ceftobiprole has also been shown to have activity against many Gram negative bacteria *in vitro* including *P. aeruginosa* (Barbour et al., 2009; Walkty et al., 2008). Both ceftobiprole and ceftaroline have been approved by the United States Food and Drug Administration for the treatment of skin and skin structure infections (Laudano, 2011; Stein et al., 2009).

Other than MRSA, *P. aeruginosa* has also been reported with evolved resistance towards the fourth generation cephalosporins. This is primarily due to AmpC β-lactamase degrading these cephalosporins. By modifying the C3 side chain of ceftazidime, a new cephalosporin CXA-101 was designed and demonstrated to be resistant to the action of *P. aeruginosa* AmpC β-lactamase (Murano et al., 2008; Toda et al., 2008). Recently, it was also shown to be effective against various β-lactam-resistant strains of *P. aeruginosa* (Giske et al., 2009; Livermore et al., 2009; Zamorano et al., 2010). Its efficacy was also tested in the presence of a β-lactamase inhibitor tazobactam against extended-spectrum β-lactamase-producing *E. coli* and *Klebsiella pneumoniae*. This study showed that the combination CXA-201 (consisting of an antibiotic and an inhibitor) was able to inhibit 96% of the test bacteria with a breakpoint of ≤ 1 mg/L (Titelman et al., 2011). Currently, both CXA-101 and CXA-201 are under Phase 2 trial (http://www.cubist.com/products/cxa_201). Selected examples of semi-synthetic cephalosporins from each generation are illustrated in Figure 4.

Industrial Synthesis of Cephalosporin Precursors

Industrial production of semi-synthetic cephalosporins began in the 1960s. The idea of producing semi-synthetic cephalosporins stemmed from the ability to synthesize a variety of semi-synthetic penicillins, using the penicillin nucleus 6-APA, with improved antimicrobial spectrum (Hamilton-Miller, 2000, 2008). However, unlike 6-APA, the cephalosporin nucleus is not readily produced in the cephalosporin producers (Rolinson and Geddes, 2007). It was clear then that the development of semi-synthetic cephalosporins as useful chemotherapeutics depended on the ability to obtain the cephalosporin nucleus.

Naturally, cephalosporin C became the initial choice to obtain the cephalosporin nucleus. By using mild acid hydrolysis, cephalosporin C could be chemically deacylated to remove the aminoadipate side chain and yield 7-ACA (Morin et al., 1962). However, the amount of 7-ACA produced was very low. By using formic acid, a yield increment from 7% to 40% could be achieved (Morin et al., 1962). This initial high yield preparation of 7-ACA allowed the synthesis of various semi-synthetic cephalosporins, i.e., thiophene-2-acetyl or furan-2-acetyl side-chains at C7β of cephem nucleus, to be tested for better efficacy than benzylpenicillin in the 1960s (Chauvette et al., 1962). The selected semi-synthetic cephalosporin was named cephalothin and was then further evaluated clinically as a broad spectrum cephalosporin (Chauvette et al., 1962). However, this chemical deacylation method required organic solvents and produced toxic chemical wastes, which are undesirable. Hence, chemical deacylation of cephalosporin C was subsequently replaced by enzymatic methods.

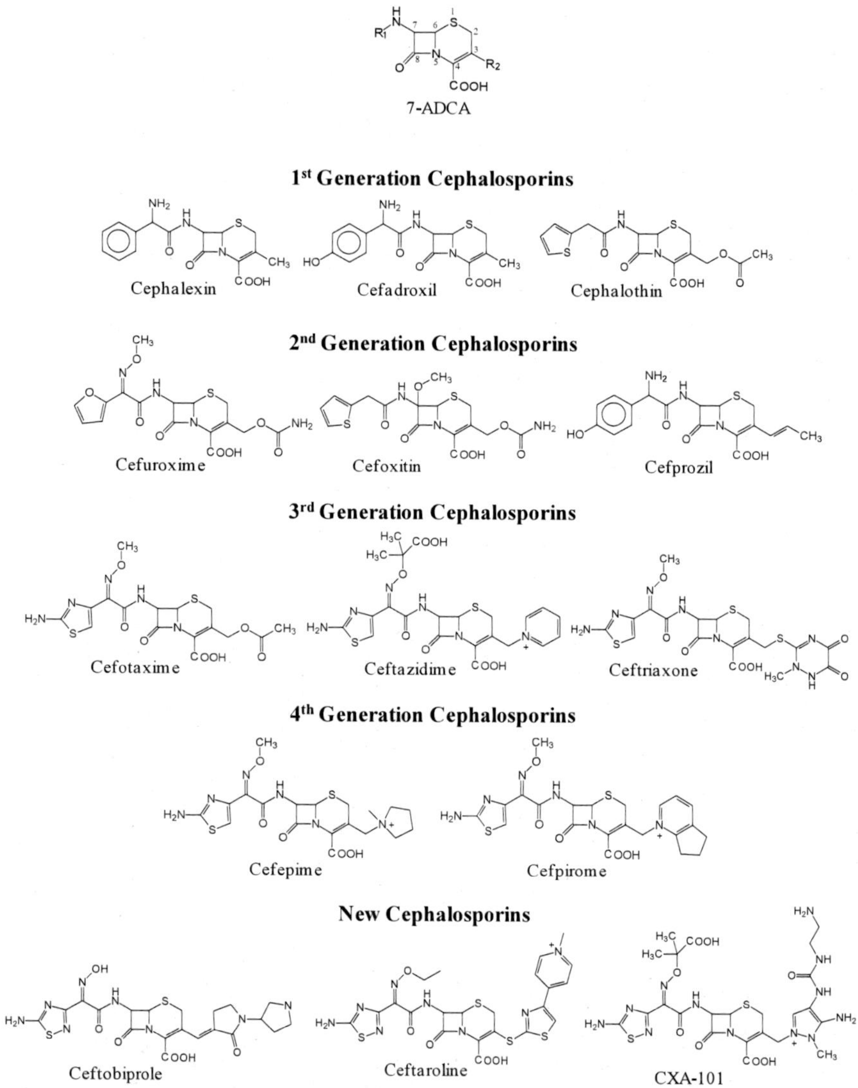

Figure 4. Chemical structures of cephalosporin nucleus 7-aminodeacetoxycephalosporanic acid (7-ADCA) and different generations of semi-synthetic cephalosporins.

Currently, there are two approaches in enzymatic deacylation. The ideal method is the direct deacylation of the aminoadipate side chain using a cephalosporin acylase. However, the natural cephalosporin acylases isolated from natural sources back then were not efficient (Barber et al., 2004). With advancement in protein engineering, the activity of cephalosporin acylase could be improved and used industrially (Oh et al., 2003; Otten et al., 2002; Pollegioni et al., 2005; Sonawane, 2006). Another approach is the oxidative deamination of cephalosporin C to keto-adipyl-7-ACA by D-amino acid oxidase (Barber et al., 2004). In this reaction, hydrogen peroxide would be produced and lead to spontaneous oxidative decarboxylation of keto-adipyl-7-ACA, which then formed glutaryl-7-ACA. Glutaryl-7-ACA is then deacylated using glutaryl acylase to finally yield 7-ACA (Figure 5). For semi-synthetic cephalosporins that require 3-vinyl substituted intermediates, e.g., cefdinir, 7-ACA has to be deacetylated to 7-aminodeacetylcephalosporanic acid (7-ADOCA) using cephalosporin acetyl hydrolase or cephalosporin acetyl esterases (Figure 5) (Barber et al., 2004; Sonawane, 2006).

Another setback with cephalosporin C production is its chemical instability in fermentation cultures of *A. chrysogenum*. This instability leading to degradation could account for approximately 40% loss of the cephalosporin C produced (Usher et al., 1988). To compete with the market of 7-ACA production, penicillin G became the preferred substrate for the production of cephalosporin nucleus because it is easily available in bulk from high yielding strain of *P. chrysogenum*. Penicillin G could be converted into a cephalosporin DAOC G by a chemical ring-expansion procedure. Subsequently, its side chain could be enzymatically deacylated to yield the cephalosporin nucleus 7-aminodeacetoxycephalosporanic acid (7-ADCA) (Figure 6). However, this chemical ring-expansion method involved multiple steps and produced toxic wastes. Hence, cheaper and greener alternatives were preferred and highly sought after.

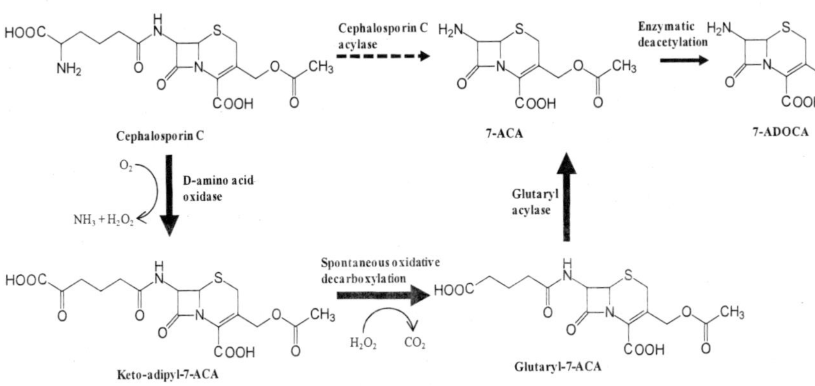

Figure 5. Industrial production of cephalosporin precursors, 7-ACA and 7-ADOCA.

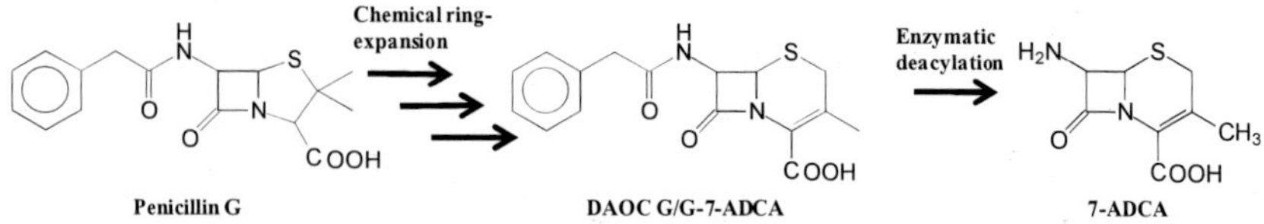

Figure 6. Industrial production of cephalosporin precursor, 7-ADCA.

GENETIC MANIPULATION OF Β-LACTAM PRODUCERS TO IMPROVE CEPHALOSPORIN PRODUCTION

Cephalosporin C per se was not an ideal antibiotic for chemotherapeutic use. Hence, its industrial production from *A. chrysogenum* was mainly to supply the cephalosporin nucleus for the production of semi-synthetics. Similar to the penicillin producers, the original cephalosporin C producer *A. chrysogenum* was not able to produce cephalosporin C on an industrial scale. Initial attempts in improving the production yield of cephalosporin C adopted similar approaches in increasing the production yield of penicillin, e.g., strain selection and classical mutagenesis (Elander, 2003). Although the yield of cephalosporin C was increased tremendously, the titre could still not match up with that of the penicillin producers (Elander, 2003).

Since both *A. chrysogenum* and *P. chrysogenum* share the same initial steps in synthesising penicillin, similar yield of cephalosporin could be produced by introducing the ring-expansion enzyme DAOCS into a high penicillin-producing *P. chrysogenum* strain to efficiently convert all the penicillin produced to cephalosporin (Table 2). However, in *P. chrysogenum*, isopenicillin N is directly converted to penicillin G or penicillin V instead of penicillin N. By introducing *S. lipmanii cef*D gene encoding isopenicillin N epimerase into *P. chrysogenum*, isopenicillin N was converted to penicillin N, which was then efficiently converted to DAOC by DAOCS encoded by *S. clavuligerus cef*E gene (Cantwell et al., 1992). This approach yielded approximately 2.5 g/L of DAOC, which was considered low compared to the titre of penicillin V produced. Furthermore, this recombinant strain continued to produce large amount of penicillin V due to the presence of IPN acyltransferase.

Subsequently, the epimerase system in *A. chrysogenum* was elucidated to be different from the bacterial epimerase system (Martin et al., 2004; Ullan et al., 2002). In *A. chrysogenum*, isopenicillin N was converted to penicillin N by a two-component protein system, which first ligates isopenicillin N to a co-enzyme A followed by epimerization of the CoA-activated aminoadipic side chain from L-configuration to D-configuration (Ullan et al., 2002). Hence, it was also likely that the earlier attempt of introducing a bacterial *cef*D gene into *P. chrysogenum* resulted in inappropriate targeting of the bacterial epimerase since *in vivo* production of penicillin is compartmentalised in *P. chrysogenum* (Evers et al., 2004; Martin et al., 2010; Ullan et al., 2007; van de Kamp et al., 1999). Therefore, subsequent effort to express *A. chrysogenum cef*D1/*cef*D2, *cef*EF and *cef*G genes in *P. chrysogenum* npe6 strain, which lacks IPN acyltransferase activity, successfully engineered *P. chrysogenum* to produce cephalosporin in large quantities (Ullan et al., 2007).

Since one of the challenges to produce cephalosporin in *P. chrysogenum* is the substrate specificity of DAOCS and DAOC/DACS for isopenicillin N, another modification approach was to feed *P. chrysogenum* with adipic acid to produce adipyl-6-APA, which could be converted to adipyl-7-ADCA and adipyl-7-ACA *in vivo* by DAOCS and DAOC/DACS, respectively. These adipyl derivatives could then be deacylated by adipyl acylase to yield 7-ADCA and 7-ACA (Crawford et al., 1995).

For the production of semi-synthetic cephalosporins, cephalosporin C has to be further processed to remove its C3 and C7β side chain. Hence, different strategies have been adapted to engineer *A. chrysogenum* to produce the cephalosporin nucleus directly (Table 2). Since naturally-occurring cephalosporin acylases have low preferences for cephalosporin C as a substrate, Isogai and co-workers (1991) transformed a gene encoding cephalosporin acylase with another gene encoding for a D-amino acid oxidase (Isogai et al., 1991). D-amino acid oxidase will deaminate cephalosporin C to keto-adipyl-7-ACA that is spontaneously decarboxylated to glutaryl-7-ACA. Then, the cephalosporin acylase will deacylate the glutaryl side chain to yield 7-ACA since studies have shown that the acylase has higher affinity for glutaryl side chain than aminoadipate side chain (Isogai et al., 1991). Another strategy was to replace the bifunctional DAOC/DACS in *A. chrysogenum* with *S. clavuligerus cef*E encoding a monofunctional DAOCS. This approach resulted in 75-80% equivalent production of DAOC compared to the parental strain producing cephalosporin C. DAOC could then be enzymatically deacylated to yield 7-ADCA using D-amino acid oxidase and glutaryl acylase (Velasco et al., 2000). Recently, Liu and co-workers (2010) also demonstrated that the earlier approach proposed by Isogai and co-workers (1991) could be realised more efficiently using a modified cephalosporin acylase with higher preference for cephalosporin C. This allowed approximately 30% of the cephalosporin C produced *in vivo* to be directly converted to 7-ACA (Liu et al., 2010).

Table 2. Genetic manipulation of β-lactam producers for the production of cephalosporin precursors

Strain	Method	Impro-vement	Remarks	Refe-rences
A. chrysogenum	Introduction of genes encoding D-amino acid oxidase and cephalosporin acylase	Production of 7-ACA	First recombinant *A. chrysogenum* strain to produce 7-ACA with a yield of approximately 150 μg/mL of 7-ACA.	Isogai et al., 1991
A. chrysogenum	Introduction of *S. clavuligerus cef*E gene	Production of DAOC instead of cephalosporin C	Production level of DAOC is nearly equivalent (75–80%) to the total β-lactams biosynthesized by the parental overproducing strain; avoids cross-contamination with other cephalosporins	Velasco et al., 2000
A. chrysogenum	Introduction of *ecs* gene, encoding cephalosporin C acylase	Production of 7-ACA	Reached 1701 μg/mL, (more than 30% of cephalosporin C produced by high-producer have been transformed into 7-ACA directly *in vivo*); highest 7-ACA production by direct fermentation ever reported	Liu et al., 2010
P. chrysogenum	Introduction of *S. clavuligerus*	Production of DAOC and lower	Maximum of 2.5 g/L of DAOC produced	Cantwell et al., 1992

	*cef*E and *Streptomyces lipmanii cef*D genes	amount of penicillin V		
P. chrysogenum	Introduction of *S. clavuligerus cef*E and feeding of adipic acid	Production of adipyl-7-ADCA	Production of 17% adipyl-7-ADCA relative to penicillin V production	Crawford et al., 1995
P. chrysogenum	Introduction of *A. chrysogenum cef*EF and *cef*G genes, and feeding of adipic acid	Production of adipyl-7-ADCA and adipyl-7-ACA	Production of 2.1-2.6% adipyl-7-ADCA, 1.4-6.7% adipyl-7-ADOCA, and 0.6% adipyl-7-ACA relative to penicillin V production	Crawford et al., 1995
P. chrysogenum npe6 (lacks IPN acyltransferase activity)	Introduction of *A. chrysogenum cef*D1 (isopenicillin N-CoA synthetase), *cef*D2 (IPN-CoA epimerase), *cef*EF, and *cef*G (DAC-acetyltransferase) genes	Production of DAC and cephalosporin C	First *P. chrysogenum* engineered strain to produce cephalosporin C; produced similar amount of β-lactam antibiotics compared to parental strain; 75% of total β-lactams produced were cephem antibiotics	Ullán et al., 2007

KEY ENZYMES IN β-LACTAM BIOSYNTHESIS

One of the strategies to improve the production of the industrial strains is to understand the biosynthetic pathways of penicillin and cephalosporin. By deciphering the key enzymes involved, increasing the amount of rate-limiting enzymes could be an approach to increase product formation and non-essential enzymes could be inhibited or removed to prevent loss of desired products. By understanding the functionalities and limitations of the key enzymes, modification could also be carried out to improve their performances for industrial applications and research purposes.

Isopenicillin N Synthase (IPNS)

IPNS plays the key role in the formation of the β-lactam ring in the penicillin biosynthesis pathway. It catalyses the cyclisation of the β-lactam ring and the ring closure of the valine moiety of ACV substrate into the five-membered thiazolidine ring. This reaction requires Fe(II) and oxygen for catalysis like most other Fe(II)-dependent oxygenases although no oxygen is incorporated into the substrate; IPNS reduces oxygen into water during the conversion of ACV to IPN (Baldwin and Abraham, 1988).

The first reported crystal structure of *A. nidulans* IPNS indicated that the tertiary structure of IPNS has eight β-strands folded to give a jelly-roll motif (also known as double-stranded β-helix fold or double Greek key motif) (Figure 7), and this was subsequently showed to be common in the NHIDOX enzymes (Clifton et al., 2006; Roach et al., 1997; Sim et al., 2003). Although IPNS bears high structural homology to other non-haem ferrous iron-dependent oxygenase and oxidases (NHIDOX), it does not require electron donors or oxidizable co-substrates, e.g., 2-oxoglutarate, which is required for the oxidative activation of the iron centre of 2-oxoglutarate-dependent dioxygenases, a sub-family of NHIDOX enzymes (Sim et al., 2003).

The reaction catalysed by IPNS is unique and had no prior synthetic organic chemistry until 1957 when formation of the β-lactam ring was accomplished in the laboratory (Sheehan and Henry-Logan, 1957). However, this multi-step chemical synthesis is laborious and economically nonviable. As such, the penicillins and cephalosporins produced industrially are still derived from naturally fermented products or via semi-synthesis with fermented β-lactams as intermediates. Nonetheless, an understanding of the biochemistry by which the β-lactam ring is formed from the tripeptide precursor continued to be sought after (Baldwin and Abraham, 1988; Baldwin and Bradley, 1990). Earlier studies on the mechanism of the IPNS reaction mainly employed enzyme kinetics, stereochemistry and a host of substrate analogs (Baldwin, 1990; Baldwin et al., 1983). These results supported a multi-step mechanism in which the formation of isopenicillin N proceeds via sequential ring closure steps to first form the β-lactam ring and then the thiazolidine ring (Baldwin and Bradley, 1990). With gene cloning, the availability of substantial amounts of highly active recombinant IPNS proteins greatly facilitated further physical and biochemical studies on the structure and function of IPNS.

Molecular studies of conserved regions of IPNS allowed the identification of the histidine and aspartate ligands in the ferrous active site. Alignment of the primary amino acid sequences of bacterial and fungal IPNSs showed that seven histidines and five aspartic acid residues are conserved throughout. These residues were changed one at a time to leucine by site-directed mutagenesis in the IPNS of *A. chrysogenum* and the recombinant proteins expressed in *E. coli* (Loke et al., 1997; Tan and Sim, 1996). Biochemical analysis revealed that His-216, Asp-218 and His-272 are essential for IPNS catalysis and these were proposed as the ligands for the active site Fe(II). Similar studies carried out with *Streptomyces jumonjinensis* IPNS led to the same conclusions (Borovok et al., 1996). Subsequently the resolved crystal structure of IPNS from *A. nidulans* confirmed these findings (Roach et al., 1997).

After the crystal structure of *A. nidulans* IPNS containing either substrate ACV or a modified substrate δ-(L-α-aminoadipyl)-L-cysteinyl-L-S-methyl-cysteine was resolved, a two-stage reaction catalysed by IPNS was deduced (Roach et al., 1997). The β-lactam ring was first formed followed by the development of a highly oxidised Fe(IV)-oxo species required for thiazolidine ring formation (Burzlaff et al., 1999). Incubation of IPNS in the presence of the substrate analogue LLL-ACV considerably decreased the turn-over of the

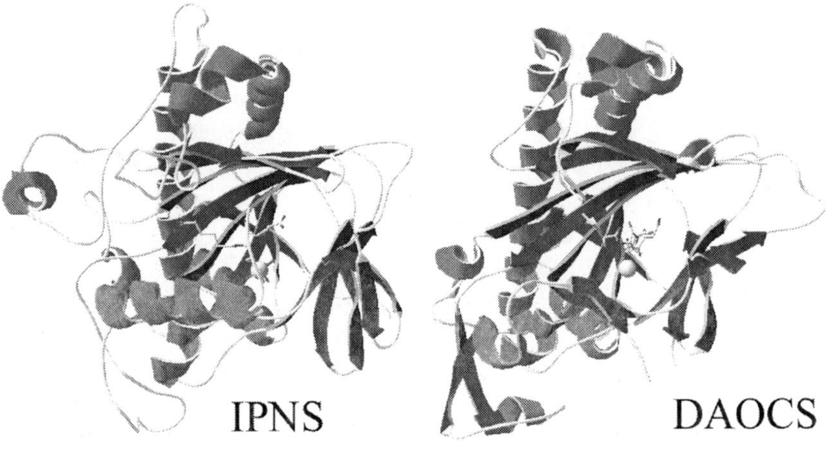

Figure 7. Crystal structures of IPNS and DAOCS illustrating similar structural fold.

enzyme via stable hydrogen bond formation within the active site. This suggests that the active site of IPNS is well suited to accommodate large hydrophobic side chains of D-amino acids in the binding pocket closed to the iron centre (Howard-Jones et al., 2005).

IPNS was earlier demonstrated to have substantial tolerance towards structural change in the tripeptide substrate (Baldwin and Bradley, 1990). A large variety of novel isopenicillin-like molecules were synthesised *in vitro* starting from ACV analogues (Baldwin and Abraham, 1988). In particular, analogues with substitution of the D-valine residue with other D-amino acid residues were well tolerated by the IPNS enzyme. Unusual bicyclic β-lactams formed include molecules with 6-, 7- and 8-membered rings fused to the β-lactam ring. Some of these unnatural β-lactams were found to act against Gram-positive and Gram-negative bacteria (Baldwin and Abraham, 1988). However, the major drawback or limitation in exploiting this enzyme for synthesizing new penicillins is the difficulty in synthesising novel tripeptide substrates. Chemical synthesis of peptide substrate is not cheap and it is probably still economical to derive the penicillin nucleus from fermentation since earlier genetic efforts in improving the industrial strains have been fruitful. Nevertheless, IPNS continues to be intriguing from a chemical and biological viewpoint, and perhaps further studies will enable its full exploitation to produce novel β-lactam structures.

Deacetoxycephalosporin C Synthase (DAOCS)

The committed step in the biosynthesis of cephem antibiotics is executed by monofunctional DAOCS in prokaryotes and bifunctional DAOC/DACS in *A. chrysogenum*. This enzymatic reaction is highly specific and has been proposed to replace the existing chemical ring-expansion of penicillin G in the industrial production of semi-synthetic cephalosporins (Andersson et al., 2001; Barber et al., 2004; Barends et al., 2004). However, the limitation in using DAOCS for enzymatic ring-expansion of penicillin G is its low catalytic activity since its native substrate is penicillin N. Hence, many studies have focused on its biochemical and structural characteristics in order to facilitate bioengineering of DAOCS for industrial application (Goo et al., 2009).

DAOCS: a Fe(II) and 2-Oxoglutarate Dependent Dioxygenase

DAOCS belongs to the protein family of 2-oxoglutarate dependent dioxygenases that utilizes Fe(II) ion and 2-oxoglutarate as a co-factor and a co-substrate, respectively, to catalyse oxidative reactions. The members in this family have a characteristic "jelly-roll" structure made up of eight β-sheets, with four β-sheets on each side of the barrel (Figure 7). The β-barrel accommodates the conserved HXD/EX_nH and RXS motifs essential for Fe(II) ion and 2-oxoglutarate binding, respectively (Hausinger, 2004). Despite the highly similar structural fold of the enzymes in this family, differences in their primary sequences define the internal architecture of the enzyme that is important for the recognition and spatial arrangement of the substrate relative to the activated iron centre in order to elicit reactions. Hence, the sequence of events from Fe(II) ion and 2-oxoglutarate binding to substrate binding and oxidation may differ slightly between enzymes. For example, while simultaneous 2-oxoglutarate and substrate binding are necessary to activate the iron centre in clavaminate synthase, binding of 2-oxoglutarate and penicillin substrates are mutually exclusive in DAOCS (Valegard et al., 2004; Zhou et al., 2001).

Biochemical Characterization of DAOCS

Deriving and optimising the enzyme reaction conditions for the detection of ring expansion activity of DAOCS were not easy tasks. It took almost two decades before an optimised *in vitro* enzyme assay could be realised. The initial reported reaction mixture contained 50 mM Tris–HCl (pH 7.4), 10 mM KCl, 10 mM $MgSO_4$, 0.6 mM ascorbic acid, 0.8 mM ATP, 0.04 mM $FeSO_4$, 0.6 mM 2-oxoglutarate, and 0.28 mM penicillin N, were used for the detection of *A. chrysogenum* DAOC/DACS ring-expansion activity (Shen et al., 1984). The order of co-factor addition to the reaction mixture was also reported to be obligatory to achieve the highest detectable activity, i.e., simultaneous addition of Fe(II) ion, ascorbic acid, and ATP followed by 2-oxoglutarate and then penicillin N (Shen et al., 1984). However, it was found that DAOCS was unable to convert penicillin G and other analogues under the optimum conditions described for penicillin N or under other varied test conditions such as temperature, duration of reaction, buffers and concentration of penicillin G (Dotzlaf and Yeh, 1989; Maeda et al., 1995). As such, DAOCS isolated from various sources was thought to have very narrow substrate specificity (Cortes et al., 1987; Dotzlaf and Yeh, 1989; Maeda et al., 1995). Subsequently, Cho and co-workers (1998) reported feasible conversion of penicillin G and 14 other penicillin N analogues to their respective cephem moieties using *S.*

clavuligerus NP1 resting cells by increasing the concentration of Fe(II) ion and 2-oxoglutarate in the reaction mixture (Cho et al., 1998). Thereafter, by omitting KCl, $MgSO_4$, dithiothreitol and ascorbic acid from the reaction mixture, Sim and Sim (2001) demonstrated that significant improvement in the conversion of penicillin G and ampicillin by recombinant *S. clavuligerus* DAOCS (scDAOCS) could be achieved (Sim and Sim, 2001). Using these improved reaction conditions, Chin and Sim (2002) eventually showed that the substrate specificity of scDAOCS was wider than previously thought (Chin and Sim, 2002).

Several assays have been used for the detection of cephalosporin formation by DAOCS. These include a hole-plate bioassay (Chin et al., 2001; Sim and Sim, 2001), thin-layer chromatography (Jensen et al., 1983), and reverse-phase high-pressure liquid chromatography (HPLC) (Chin et al., 2001; Fernandez et al., 1999; Jensen et al., 1983). A hole-plate bioassay relies on the use of agar plate seeded with β-lactam-sensitive *E. coli* and penicillinase, which will degrade any remaining penicillin substrate. A well is made in the centre of the agar plate followed by loading of enzyme reaction mixture. The antibiotic compounds present in the mixture will diffuse into the agar and inhibit the growth of *E. coli*, resulting in a zone of growth inhibition, which reflects the concentration of cephalosporin present. Thin-layer chromatography uses a cellulose paper as the solid phase and *n*-butanol-acetic acid-water as the mobile phase to separate penicillin substrate and cephalosporin product based on their solubility differences in the mobile phase. After development, the paper is dried and placed face-down on an agar slab containing β-lactam-sensitive *E. coli* and penicillinase to detect for growth inhibition. Reverse-phase HPLC is similar to thin-layer chromatography except that it uses a C_{18} column instead of cellulose paper and the separation of compounds is under high pressure. The elution of respective penicillin substrate and cephalosporin product is monitored using a spectrophotometer, which will read out the level of absorption at a specific wavelength of light against retention time. However, these assays are laborious, cumbersome, and yield limited data points. Furthermore, a large number of samples are required for kinetic studies. Subsequently, a spectrophotometric assay with continuous detection of penicillin conversion by DAOCS was devised by Dubus and co-workers (2001) and has allowed a more precise determination of DAOCS activity. They showed that scDAOCS has a K_m value of 0.89 mM and a k_{cat} value of 0.079 s^{-1} for penicillin G conversion, and a K_m value of 4.86 mM and a k_{cat} value of 0.120 s^{-1} for ampicillin conversion (Dubus et al., 2001).

Structural and Mechanistic Studies of DAOCS

The primary interest in DAOCS is its ability to ring-expand penicillin to cephalosporin. This key difference in the dihydrothiazine ring confers higher stability of the β-lactam ring to penicillinase. The first crystal structure of DAOCS from *S. clavuligerus* provided an important view of the tertiary folding of the enzyme and physical evidence of the residues involved in the coordination of co-factor ferrous iron and co-substrate 2-oxoglutarate (Valegard et al., 1998). This resolved structure revealed highly similar secondary and tertiary foldings similar to that of IPNS. This study also revealed that scDAOCS apoenzyme exists as a trimeric unit, where the C-terminal residues 308–311 of one molecule penetrate the active site of its neighbouring molecule in a cyclical manner. In the presence of Fe(II) and 2-oxoglutarate, scDAOCS dissociates into monomers. These observations were supported by gel filtration and dynamic laser light scattering studies which revealed the existence of an equilibrium between the monomeric and trimeric scDAOCS in the absence of Fe(II). Upon addition of Fe(II) and 2-oxoglutarate, there was almost a complete shift towards the monomeric form, suggesting that the monomeric scDAOCS is the catalytically active form facilitating the oxidative reaction (Lloyd et al., 1999). However, the possible existence and significance of this trimeric state *in vivo* remain to be elucidated.

The first crystal structure of DAOCS, which lacked a bound substrate, had limited the understanding of the catalytic mechanism of scDAOCS. The stable native trimeric structure of scDAOCS made crystallisation of an enzyme–substrate complex difficult, as the active site is partially occupied by the C-terminal tail of the neighbouring enzyme. Truncation of the C-terminal tail by six residues helped to prevent trimer formation but this approach resulted in a loss of enzymatic activity. By attaching a His-tag at the N-terminus of scDAOCS, Oster and co-workers (2004) reported that the trimer formation was disrupted and the C-terminal arm of the apoenzyme was liberated (Oster et al., 2004). However, biochemical and structural analyses revealed significant changes in the substrate-binding affinity and orientation of the penicillin nucleus in the His-tagged scDAOCS–substrate complex structure compared with those in the native enzyme–substrate structure published by the same group (Valegard et al., 2004). These discrepancies suggest caution when selecting the appropriate structural template for subsequent studies. Nevertheless, in both types of structure, the C-terminal residues 308-311 remained disordered in the enzyme–substrate complex, implying a degree of flexibility that may be crucial for catalytic activity, as supported by mutagenesis studies (Chin et al., 2001; Chin and Sim, 2002; Lee et al., 2001; Lee et al., 2002).

The elucidation of scDAOCS structures with Fe(II), 2-oxoglutarate, and penicillins was achieved by soaking the crystals of apo-scDAOCS with Fe(II) followed by soaking in solution containing 2-oxogluatrate and either penicillin G or ampicillin. The structural information obtained provides the basis of the current "booby-trap" mechanism hypothesis (Valegard et al., 2004). The postulated mechanism involves initial activation of the iron centre by 2-oxoglutarate binding followed by oxidative decarboxylation of 2-oxoglutarate to generate oxidizing ferryl species with succinate and carbon dioxide as by-products. Succinate remains bound and stabilises the reactive iron species until the penicillin substrate enters the active site. Binding of penicillin to the reactive iron species via the sulphur group subsequently expels succinate and leads to oxidative ring expansion of the penicillin substrate. This study also illustrated sharing of overlapping binding sites for substrate and co-substrate, which implies mutual exclusivity in substrate and co-substrate binding. This requirement of an ordered sequential binding for catalysis supported earlier findings on the importance of the order of adding reagents in the enzyme (Shen et al., 1984). Furthermore, the difference in the orientation of penicillin substrates in DAOCS complexes formed with or without 2-oxoglutarate suggests that 2-oxoglutarate is necessary to influence the orientation of penicillin binding. This is in accordance to the requirement of 2-oxoglutarate to activate the iron centre in order to elicit oxidative ring-expansion. Hence, the orientation of the penicillin substrate in DAOCS formed in the presence of 2-oxoglutarate, which also resembled the orientation of product DAOC co-crystallised with DAOCS, advocated a productive ring-expansion of penicillin (Valegard et al., 2004).

Other than providing insights to the mechanism of catalysis, the crystal structures of DAOCS with penicillins have also provided important structural information to enzyme-substrate interaction. Some of these residues have the potential to improve the enzymatic activity of DAOCS by mutational alterations, which will be discussed in the next section.

Modification of DAOCS to Improve its Substrate Binding and Catalytic Activity

The bottleneck in the direct application of DAOCS for industrial ring-expansion of penicillin G is the substrate specificity of DAOCS. Although scDAOCS was the most active among the orthologues tested using penicillin

G (Chin et al., 2003) and has broad substrate specificity (Goo et al., 2009), its low substrate affinity for penicillin G limited its potential application (Wei et al., 2003). Kinetic studies of scDAOCS using HPLC showed that the K_m and k_{cat} values for penicillin G conversion were 2.58 mM and 0.0453 s^{-1}, and for penicillin N conversion were 0.014 mM and 0.307 s^{-1}, respectively (Wei et al., 2003). This nearly 1200-fold lower catalytic efficiency of scDAOCS for penicillin G compared to penicillin N drove the efforts towards understanding the structure–function relationship of scDAOCS and improving its catalytic efficiency in converting penicillin G.

Recombinant DAOCSs

The first reported attempt to improve the activity of DAOCS was probably the construction of a hybrid from *S. clavuligerus* and *N. lactamdurans* DAOCSs using *in vivo* homologous recombination (Adrio and Demain, 2002; Adrio et al., 2002). This technique exploited the natural recombination of homologous sequences in a host organism to generate a recombinant enzyme. *Streptomyces lividans* strain 1326 was used as a host and a tyrosinase *mel*C gene from *Streptomyces glaucescens* was used as a reporter gene, which encodes the ability to secrete the black pigment melanin. Recombinants were screened for by the loss of pigment production in the host organism due to the homologous recombination of the two *cef*E genes flanking the *mel*C gene in a pJA680 vector. A recombinant strain W25 was generated and found to have higher penicillin G-conversion activity than the *S. clavuligerus* NP1 strain (Adrio and Demain, 2002; Adrio et al., 2002). Further analyses revealed that this recombinant strain preferred a lower reaction temperature (<30°C), a lower Fe(II) concentration, and a higher level of 2-oxoglutarate for penicillin G conversion than *S. clavuligerus* NP1 strain (Gao et al., 2003).

Another hybrid DAOCS was also generated but the method used was an *in vitro* recombination method, i.e., gene shuffling, and seven gene sequences were used, which were *cef*E genes from *S. clavuligerus* ATCC 27064, *N. lactamdurans* ATCC 27382, *S. jumonjinensis* ATCC 29864, *S. ambofaciens* 102SH3 and *S. chartreusis* 29SA4, *cef*F genes from *S. clavuligerus* and isolate 65PH1, and *cef*EF gene from *A. chrysogenum* ATCC 11550. These shuffled genes were heterologously expressed in *E. coli* cells and the transformants were screened for penicillin G conversion activity. After two rounds of gene shuffling and screening, a clone known as FF8 was identified, which exhibited no substrate inhibition and has a 118-fold higher k_{cat}/K_m value when compared with scDAOCS (Hsu et al., 2004). Comparison of the amino acid sequence of FF8 with that of scDAOCS revealed that only two of the 35 substitutions, i.e., residues M73 and I305, are at positions known to have direct interactions with the substrate. This suggests that the substrate specificity of DAOCS is not solely governed by the substrate binding residues and amino acid residues outside the substrate-binding site may also affect the substrate specificity of the enzyme.

Other methods have also been used to improve the ring-expansion activity of DAOCS. However, the extent of improvement achieved was not as great as what was achieved for recombination method. Nevertheless, these studies have provided important clues for rational engineering of DAOCS to further enhance its catalytic functions. Chemical mutagenesis using hydroxylamine was attempted and the mutant library generated was first screened by a simple high throughput paper disk-based bioassay followed by a second screening method using thin-layer chromatography with radio-labeled penicillin G as a substrate. From 5,500 clones screened, three point mutations, G79E, V275I and C281Y, that led to enhanced ring-expansion of penicillin G were identified (Wei et al., 2003). Since these three sites are located away from the substrate binding pocket, the improvement achieved could likely be due to a global effect of these mutations on the conformation of the active site (Wei et al., 2003).

Another method attempted by the same group of researchers was the combination of error-prone PCR and gene shuffling (Wei et al., 2005). Error-prone PCR exploits the altered fidelity of DNA polymerase in the presence of manganese ions to incorporate mismatch nucleotides during PCR amplification of the gene of interest. A ratio of Mg (II) ions to Mn (II) ions of 12:1 in the error-prone PCR reaction was used to induce mutations. Out of 6,364 clones screened, six types of point mutations (M73T, T91A, M188V, M188I, H244Q, G300V) and six types of double mutations (D53H/C281Y, M73T/P145L, A106T/C155Y, A106T/M188V, D107G/L277Q, Y184H/C281R), which presented higher catalytic activities, were identified. Subsequently, single mutations based on the double mutations were constructed to assess the effect of each substitution. Among these mutations, the L277Q mutation resulted in highest catalytic efficiency for penicillin G conversion (six-fold increment), whereas the M73T, G79E, T91A, Y184H, M188V, and H244Q mutations resulted in approximately two to five-fold increment in k_{cat}/K_m values when compared with the wild-type enzyme. Other mutations either had activities similar to that of the wild-type enzyme or exhibited substrate inhibition (Wei et al., 2005).

Table 3. Kinetic parameters of mutant DAOCS in penicillin G conversion (Hsu et al., 2004; Wei et al., 2005; Wei et al., 2003)

Mutations	K_m (mM)	k_{cat} (s^{-1})	k_{cat}/K_m ($M^{-1}s^{-1}$)	Relative k_{cat}/K_m
Wild-type	2.58	0.0453	18	1
M73T	0.74	0.0627	85	4.7
G79E	0.75	0.0315	42	2.3
T91A	1.36	0.0522	38	2.1
C155Y	1.76	0.0476	27	1.5
Y184H	0.95	0.0798	84	4.7
M188V	0.76	0.0456	60	3.3
H244Q	1.39	0.0698	50	2.8
V275I	1.68	0.0502	30	1.7
L277Q	1.02	0.1036	102	5.7
C281Y	0.68	0.0744	109	6.1
N304K	0.22	0.0564	256	14.2
I305L	0.66	0.0759	115	6.4
I305M	0.75	0.1452	194	10.8
V275I/I305M	0.25	0.1458	583	32.4
C155Y/Y184H/V275I/C281Y	0.19	0.1398	736	40.9
FF8	0.014	0.0297	2,121	117.8

Gene shuffling of these single mutants, including those reported previously, i.e., G79E, V275I, C281Y, N304K, I305L, and I305M mutants (Wei et al., 2003), was then conducted to generate random combinations. Approximately 10,000 clones were screened and only one quaternary mutant (C155Y/Y184H/V275I/C281Y) with enhanced activity and without substrate inhibition was subjected to kinetic analysis. This mutant was about 40-fold more efficient than the wild-type enzyme and about seven-fold higher than the C281Y single mutant when comparing their k_{cat}/K_m values for penicillin G conversion. However, this mutant still cannot match up with the catalytic activity of clone FF8 (Table 3).

Rational Engineering of DAOCS

Rational engineering approach for improving the substrate binding affinity of DAOCS has been challenging initially due to the lack of a three-dimensional structure of enzyme-substrate complex. Although the three-dimensional structure of the DAOCS-Fe(II)-2-oxoglutarate complex was resolved in the late 1990s (Lloyd et al., 1999; Valegard et al., 1998), the lack of bound penicillin substrate hindered many subsequent studies. However, it was recognized that the crystal structure of DAOCS resembles that of IPNS (Roach et al., 1997; Valegard et al., 1998). By superimposing the crystal structure of the DAOCS-Fe(II)-2-oxoglutarate complex on that of the IPNS-Fe(II)-ACV complex, it was possible to predict the substrate binding pocket of scDAOCS since the substrates of both enzymes are structurally similar (Chin et al., 2001; Valegard et al., 1998). Based on the close proximity to the L-α-aminoadipyl moiety of ACV, amino acid residues in the predicted substrate binding pocket of scDAOCS were assessed and selected for mutational studies (Chin et al., 2001). This study showed that N304L mutation increased the conversion activity of scDAOCS by approximately two-fold for penicillin G, ampicillin and amoxicillin relative to the wild-type enzyme. Subsequent studies showed that N304L, R306L and R307L single mutations were able to improve the conversion activity of scDAOCS to different extents, ranging from 1.2- to four-fold, for seven penicillin substrates, namely penicillin G, ampicillin, amoxicillin, metampicillin, carbenicillin, penicillin V and phenethicillin (Chin and Sim, 2002). Thereafter, Wei and co-workers (2003) also adopted the same approach and reported that the N304K, I305L, and I305M mutations were able to increase the conversion activity of scDAOCS by about two- to six-fold using penicillin G as a substrate. Similarly, mutagenesis studies were done on *A. chrysogenum* DAOCS/DACS to probe its amenability to modification to improve its catalytic activity towards various penicillin substrates (Wu et al., 2005). N305L mutation of DAOC/DACS, which is the corresponding N304L mutation of scDAOCS, was also demonstrated to increase the conversion activity of DAOC/DACS using penicillin G, ampicillin and carbenicillin as substrates (Wu et al., 2005). Various structural studies, together with these mutational studies, suggested the possible involvement of C-terminus residues of DAOCS in guiding the substrate into the substrate binding pocket (Lee et al., 2001; Valegard et al., 1998). However, the rationale behind these mutations could not be fully comprehended in the absence of a crystal structure of DAOCS complexed with a penicillin substrate.

In order to elucidate the molecular basis of substrate interaction by residue N304, Chin and co-workers (2004) performed a complete library of amino acid alterations at this position and found that mutation to basic residues (arginine and lysine) has better effects than mutation to aliphatic residues (alanine, leucine, and methionine) in the conversion of seven hydrophobic penicillins (Chin et al., 2004). Supported by the resolution of a crystal structure of scDAOCS complex with penicillin G and ampicillin, it was realized that the side chain of asparagine at position 304 aligns with that of the penicillin substrate (Valegard et al., 2004). Hence, the observed improvement is probably due to better hydrophobic interaction between the side chains of arginine and lysine and that of the penicillin substrate, leading to a lower K_m value (Goo et al., 2008b).

A similar approach to the N304 mutation studies was adopted to further understand the role of residue R306 in the conversion activity of scDAOCS (Goo et al., 2008a). The crystal structure of scDAOCS-Fe(II)-penicillin complex revealed that the guanidinium side chain of residue R306 is located away from the catalytic site and interacts with a hydrophobic cleft formed by residues Y184, L186, W297, I298, and V303 (Goo et al., 2008a; Valegard et al., 2004). It was clear that the improved conversion activity conferred by R306L mutation was not because of direct interaction with the substrate. From the mutational studies of R306, it was deduced that the length and hydrophobicity of the amino acid side chain at this position are important in influencing the function of the enzyme (Goo et al., 2008a). The loss of enzyme activity by R306P mutation indicated that the flexibility of scDAOCS C-terminus is crucial for its catalytic property. Replacement with polar and charged residues resulted in a tremendous decrease in the conversion activity of scDAOCS, which retained approximately 40% of the wild-type activity or lower, for all substrates tested. This supported the postulation that polar and charged residues are detrimental to this site, probably because they perturb the integrity of the hydrophobic cleft. Although both arginine and lysine residues are positively-charged amino acids, being buried in a hydrophobic environment would reduced the pK_a of their side chains (Guillen Schlippe and Hedstrom, 2005). This would render the amino groups in the side chains of both residues unprotonated and uncharged. Consequently, the integrity at this region might be preserved. Replacement with bulky aromatic residues like phenylalanine, tyrosine and tryptophan, was also not ideal probably to structural hindrance. Replacement of R306 with amino acids structurally similar to leucine, i.e., isoleucine and methionine, was able to improve the enzyme activity via hydrophobic interaction with the surrounding residues. As reflected by their K_m values for ampicillin conversion, the improvements were due to enhancement of the substrate-binding affinity of the enzyme by these mutations (Goo et al., 2008a). Interestingly, a study on residue R307 of DAOC/DACS, which is the corresponding residue of scDAOCS R306, showed different modification effects upon the same type of substitution, i.e., from arginine residue to leucine. R306L mutation in scDAOCS resulted in modest increment in the ring-expansion activity for penicillin G (Chin and Sim, 2002; Goo et al., 2008a) but not for R307L mutation in DAOC/DACS (Wu et al., 2005). This could be due to the additional 20 amino acid residues at the C-terminal of DAOC/DACS influencing the interaction of the amino acid residue at position 307. This discrepancy will only be clearer when the tertiary structure of DAOC/DACS has been resolved.

Although the role of residue R307 and the mechanism behind R307L mutation in improving the catalytic activity of scDAOCS has yet to be elucidated, further studies on the corresponding residue R308 of DAOC/DACS were done to assess the effects of all possible substitutions in the substrate specificity of DAOC/DACS (Wu et al., 2011). Interestingly, both R308I and R308L mutations resulted in approximately 12- and eight-fold increment in k_{cat} values, respectively, although they affected the substrate binding affinity of DAOC/DACS for penicillin G. Nonetheless, the resulting catalytic efficiencies of these two mutants were about twice that of the wild-type enzyme indicating an overall improvement of the catalytic activity of DAOC/DACS (Wu et al., 2011).

Although this approach of using a complete library of amino acid replacement mutants is laborious and time-consuming, the effects of all possible mutations at that position of the protein could be determined and the ideal amino acid residue that improves the function of the enzyme would also be revealed. This would then allow rational selection of useful mutations for combinatorial studies since more than one mutation of the enzyme is often required to alter the property of an enzyme towards a desirable one. When two or more mutations are combined in a single protein, there could be various effects (Mildvan et al., 1992; Wells, 1990). This is probably the reason why recombinant DAOCSs obtained so far were more superior to those obtained via rational approach in terms of their catalytic properties. The combination of recombination and high through-put screening for recombinant enzymes with higher ring-expansion activity allowed the selection of superior catalysis. However, the rationale behind the improvement obtained from such approach often could not be easily comprehended. Therefore, it is necessary to critically assess the effects of each combination on its own merits. Hence, various combinations of double mutation based on V275I, C281Y, and C-terminal mutations (N304A, N304K, N304L, N304M, N304R, I305M, R306L and R307L) were generated and evaluated (Goo et al., 2008b). Kinetic analyses using ampicillin as a substrate showed that the k_{cat}/K_m values of all double-mutant scDAOCSs were higher than that of the wild-type enzyme using ampicillin as a substrate. Two double mutants, C281Y/N304R and

C281Y/I305M, showed the greatest improvement in the catalytic efficiency of scDAOCS (101 and 68-fold increments, respectively). This study also showed that N304K and N304R mutations tend to improve the substrate-binding affinity of the enzyme whereas I305M mutation tends to increase the turnover rate (Goo et al., 2008b).

FUTURE DEVELOPMENT OF β-LACTAM ANTIBIOTICS

The serendipitous discovery of penicillin has revolutionised the treatment of infectious diseases but it has also resulted in the selection of resistant pathogens. Other than the need for broad spectrum penicillin, the loss of efficacy of penicillin to resistant pathogens has called for the development of β-lactamase-resistant penicillin in the 1960s. The recognition of cephalosporin being more penicillinase-resistant and the availability of two alkyl sites amenable to modification have led to the development of a plethora of semi-synthetic cephalosporins with diverse antimicrobial spectra and resistant to various classes of β-lactamases. However, the vicious cycle of development of new antibiotics followed by emergence of resistance to them has dampened many pharmaceutical companies to discover new antibiotics or develop better antibiotics. With the large number of different antibiotics being discovered and developed over the last seven decades, the cost of searching or developing new antibiotics has become more demanding than ever. Hence, it is necessary to derive more efficient method in drug screening, and deeper understanding of structure-activity relationship of antibiotics with respect to the drug targets and incorporating pharmacokinetics at the same time in order to obtain a truly efficacious drug without developing resistance, if possible.

Both cephamycin and cephabacin have been demonstrated to be relatively more resistant to the action of some β-lactamases than cephalosporin C and yet, semi-synthetics developed based on these structures are limited or none in the case of cephabacin. This is probably due to the similar theme of structural modification at C3 and C7β positions for these cephem antibiotics. Since the presence of a functional group at the C7α position of cephamycin and cephabacin probably account for the increased stability to β-lactamases, the use of cephamycin and cephabacin nuclei for future development of cephalosporin semi-synthetics may confer higher resistance to the highly active β-lactamases.

From the studies of cephabacin biosynthesis, the enzymes involved in the synthesis of the oligopeptide side chain have been determined to be a NRPS-PKS complex (Demirev et al., 2006). NRPS is an important enzyme for the synthesis of oligopeptides in contrast to ribosomes, which synthesise polypeptides. Variation in the substrate specificity of each module in the NRPS will allow the fusion of various amino acids into diverse combination of oligopeptides (Strieker et al., 2010). PKS is an enzyme containing multiple domains catalysing different reactions. These reactions comprise of keto-synthase, acyltransferase, keto-reductase, acyl carrier protein and thioesterase activities, which fused short carbon molecules into long chain polyketide similar to fatty acid biosynthesis (Khosla, 2009). This highlighted the possibility of generating a diverse group of oligopeptides to be attached to the C3 position of cephalosporin by varying the modules in NRPS-PKS complex. This will generate a range of new cephalosporins and allow testing of their efficacies against highly resistant pathogens like the ESKAPE pathogens (Boucher et al., 2009). Since penicillin and cephalosporin target PBPs by mimicking their native substrates, developing cephalosporin with oligopeptide moiety may probably generate a better fit to the substrate binding pocket of PBPs (Goo and Sim, 2011). Furthermore, many NRPSs and PKSs have been involved in the biosynthesis of medically important compounds (Felnagle et al., 2008). Hence, it is perhaps feasible to synthesize novel structure with antibiotic activity by exploiting this NRPS-PKS complex involved in cephabacin biosynthesis and fused with cephem structure (Menzella and Reeves, 2007).

Other than producing more efficacious antibiotic, it is also important to understand the origin of resistance, dissemination of resistant factor, evolution of resistance, proper administration of antibiotic and regulation of antibiotic use instead of easy availability in the pharmacy. It is perhaps inevitable that for very new antibiotic develop, new resistance will emerge since it is a selection process. However, studies have showed the relevance of antibiotic concentration in resistance development (Gould and MacKenzie, 2002). More importantly, recent studies on the effects of sub-inhibitory antibiotic concentration revealed various physiological changes that have important implications to resistance development, including mutation frequency (Davies et al., 2006; Henderson-Begg et al., 2006). With better and advanced understanding of resistance development, the beneficial effects of antibiotics will be truly appreciated and their use will be justifiably regulated. In turn, this may result in reduction of the rate of emergence of resistant pathogens and hopefully provide better control over the spread of infectious diseases.

REFERENCES

Abraham, E.P. and Newton, G.G. (1961) 'The structure of cephalosporin C', *Biochem J*, Vol. 79, pp. 377-393.

Adrio, J.L. and Demain, A.L. (2002) 'Improvements in the formation of cephalosporins from penicillin G and other penicillins by bioconversion', *Organic Process Research & Development*, Vol. 6, No. 4, pp. 427-433.

Adrio, J.L., Hintermann, G.A., Demain, A.L. and Piret, J.M. (2002) 'Construction of hybrid bacterial deacetoxycephalosporin C synthases (expandases) by in vivo homeologous recombination', *Enzyme and Microbial Technology*, Vol. 31, No. 7, pp. 932-940.

Alvarez, E., Meesschaert, B., Montenegro, E., Gutierrez, S., Diez, B., Barredo, J.L. and Martin, J.F. (1993) 'The isopenicillin-N acyltransferase of *Penicillium chrysogenum* has isopenicillin N amidohydrolase, 6-aminopenicillanic acid acyltransferase and penicillin amidase activities, all of which are encoded by the single *pen*DE gene', *Eur J Biochem*, Vol. 215, No. 2, pp. 323-332.

Andersson, I., van Scheltinga, A.C.T. and Valegard, K. (2001) 'Towards new beta-lactam antibiotics', *Cellular and Molecular Life Sciences*, Vol. 58, No. 12-13, pp. 1897-1906.

Aoki, H., Sakai, H., Kohsaka, M., Konomi, T. and Hosoda, J. (1976) 'Nocardicin A, a new monocyclic beta-lactam antibiotic. I. Discovery, isolation and characterization', *J Antibiot (Tokyo)*, Vol. 29, No. 5, pp. 492-500.

Appelbaum, P.C. (2006) 'MRSA - the tip of the iceberg', *Clin Microbiol Infect*, Vol. 12 Suppl 2, pp. 3-10.

Baldwin, J.E. (1990) 'The biosynthesis of penicillins and cephalosporins', *Journal of Heterocyclic Chemistry*, Vol. 27, No. 1, pp. 71-78.

Baldwin, J.E. and Abraham, E. (1988) 'The biosynthesis of penicillins and cephalosporins', *Nat Prod Rep*, Vol. 5, No. 2, pp. 129-145.

Baldwin, J.E., Abraham, E.P., Adlington, R.M., Chakravarti, B., Derome, A.E., Murphy, J.A., Field, L.D., Green, N.B., Ting, H.H. and Usher, J.J. (1983) 'Penicillin biosynthesis - dual pathways from a modified substrate', *Journal of the Chemical Society-Chemical Communications*, No. 22, pp. 1317-1319.

Baldwin, J.E. and Bradley, M. (1990) 'Isopenicillin N synthase - mechanistic studies', *Chemical Reviews*, Vol. 90, No. 7, pp. 1079-1088.

Barber, M.S., Giesecke, U., Reichert, A. and Minas, W. (2004) 'Industrial enzymatic production of cephalosporin-based beta-lactams': *Molecular*

Biotechnology of Fungal Beta-Lactam Antibiotics and Related Peptide Synthetases, pp. 179-215.

Barbour, A., Schmidt, S., Rand, K.H. and Derendorf, H. (2009) 'Ceftobiprole: a novel cephalosporin with activity against Gram-positive and Gram-negative pathogens, including methicillin-resistant *Staphylococcus aureus* (MRSA)', *Int J Antimicrob Agents*, Vol. 34, No. 1, pp. 1-7.

Barends, T.R., Yoshida, H. and Dijkstra, B.W. (2004) 'Three-dimensional structures of enzymes useful for beta-lactam antibiotic production', *Curr Opin Biotechnol*, Vol. 15, No. 4, pp. 356-363.

Barredo, J.L., van Solingen, P., Diez, B., Alvarez, E., Cantoral, J.M., Kattevilder, A., Smaal, E.B., Groenen, M.A., Veenstra, A.E. and Martin, J.F. (1989) 'Cloning and characterization of the acyl-coenzyme A: 6-aminopenicillanic-acid-acyltransferase gene of *Penicillium chrysogenum*', *Gene*, Vol. 83, No. 2, pp. 291-300.

Borovok, I., Landman, O., Kreisberg-Zakarin, R., Aharonowitz, Y. and Cohen, G. (1996) 'Ferrous active site of isopenicillin N synthase: genetic and sequence analysis of the endogenous ligands', *Biochemistry*, Vol. 35, No. 6, pp. 1981-1987.

Boucher, H.W., Talbot, G.H., Bradley, J.S., Edwards, J.E., Gilbert, D., Rice, L.B., Scheld, M., Spellberg, B. and Bartlett, J. (2009) 'Bad bugs, no drugs: no ESKAPE! An update from the Infectious Diseases Society of America', *Clin Infect Dis*, Vol. 48, No. 1, pp. 1-12.

Brakhage, A.A., Al-Abdallah, Q., Tuncher, A. and Sprote, P. (2005) 'Evolution of beta-lactam biosynthesis genes and recruitment of trans-acting factors', *Phytochemistry*, Vol. 66, No. 11, pp. 1200-1210.

Burzlaff, N.I., Rutledge, P.J., Clifton, I.J., Hensgens, C.M., Pickford, M., Adlington, R.M., Roach, P.L. and Baldwin, J.E. (1999) 'The reaction cycle of isopenicillin N synthase observed by X-ray diffraction', *Nature*, Vol. 401, No. 6754, pp. 721-724.

Bush, L.M. and Johnson, C.C. (2000) 'Ureidopenicillins and beta-lactam/beta-lactamase inhibitor combinations', *Infect Dis Clin North Am*, Vol. 14, No. 2, pp. 409-433, ix.

Campos Muñiz, C., Zelaya, T.E.C., Esquivel, G.R. and Fernández, F.J. (2007) 'Penicillin and cephalosporin production: a historical perspective', *Rev Latinoam Microbiol*, Vol. 49, No. 3-4, pp. 88-98.

Cantwell, C., Beckmann, R., Whiteman, P., Queener, S.W. and Abraham, E.P. (1992) 'Isolation of deacetoxycephalosporin C from fermentation broths of *Penicillium chrysogenum* transformants: construction of a new fungal biosynthetic pathway', *Proc Biol Sci*, Vol. 248, No. 1323, pp. 283-289.

Casqueiro, J., Gutierrez, S., Banuelos, O., Hijarrubia, M.J. and Martin, J.F. (1999) 'Gene targeting in *Penicillium chrysogenum*: disruption of the *lys*2 gene leads to penicillin overproduction', *J Bacteriol*, Vol. 181, No. 4, pp. 1181-1188.

Castro, J.M., Liras, P., Laiz, L., Cortes, J. and Martin, J.F. (1988) 'Purification and characterization of the isopenicillin N synthase of *Streptomyces lactamdurans*', *J Gen Microbiol*, Vol. 134, No. 1, pp. 133-141.

Chain, E. (1972) 'Thirty years of penicillin therapy', *J R Coll Physicians Lond*, Vol. 6, No. 2, pp. 103-131.

Chauvette, R.R., Spencer, J.L., Mueller, R.A., Roeske, R.W., Morin, R.B., Ryan, C.W., Jackson, B.G., Pioch, R.P., Flynn, E.H., Vanheyningen, E. and Lavagnino, E.R. (1962) 'Chemistry of cephalosporin antibiotics .2. Preparation of a new class of antibiotics and relation of structure to activity', *Journal of the American Chemical Society*, Vol. 84, No. 17, pp. 3401.

Chin, H.S., Goo, K.S. and Sim, T.S. (2004) 'A complete library of amino acid alterations at N304 in *Streptomyces clavuligerus* deacetoxycephalosporin C synthase elucidates the basis for enhanced penicillin analogue conversion', *Applied and Environmental Microbiology*, Vol. 70, No. 1, pp. 607-609.

Chin, H.S., Sim, J., Seah, K.I. and Sim, T.S. (2003) 'Deacetoxycephalosporin C synthase isozymes exhibit diverse catalytic activity and substrate specificity', *Fems Microbiology Letters*, Vol. 218, No. 2, pp. 251-257.

Chin, H.S., Sim, J. and Sim, T.S. (2001) 'Mutation of N304 to leucine in *Streptomyces clavuligerus* deacetoxycephalosporin C synthase creates an enzyme with increased penicillin analogue conversion', *Biochemical and Biophysical Research Communications*, Vol. 287, No. 2, pp. 507-513.

Chin, H.S. and Sim, T.S. (2002) 'C-terminus modification of *Streptomyces clavuligerus* deacetoxycephalosporin C synthase improves catalysis with an expanded substrate specificity', *Biochem Biophys Res Commun*, Vol. 295, No. 1, pp. 55-61.

Cho, H., Adrio, J.L., Luengo, J.M., Wolfe, S., Ocran, S., Hintermann, G., Piret, J.M. and Demain, A.L. (1998) 'Elucidation of conditions allowing conversion of penicillin G and other penicillins to deacetoxycephalosporins by resting cells and extracts of *Streptomyces clavuligerus* NP1', *Proceedings of the National Academy of Sciences of the United States of America*, Vol. 95, No. 20, pp. 11544-11548.

Clifton, I.J., McDonough, M.A., Ehrismann, D., Kershaw, N.J., Granatino, N. and Schofield, C.J. (2006) 'Structural studies on 2-oxoglutarate oxygenases and related double-stranded beta-helix fold proteins', *Journal of Inorganic Biochemistry*, Vol. 100, No. 4, pp. 644-669.

Coque, J.J., Enguita, F.J., Martin, J.F. and Liras, P. (1995a) 'A two-protein component 7 alpha-cephem-methoxylase encoded by two genes of the cephamycin C cluster converts cephalosporin C to 7-methoxycephalosporin C', *J Bacteriol*, Vol. 177, No. 8, pp. 2230-2235.

Coque, J.J., Martin, J.F., Calzada, J.G. and Liras, P. (1991) 'The cephamycin biosynthetic genes *pcb*AB, encoding a large multidomain peptide synthetase, and *pcb*C of *Nocardia lactamdurans* are clustered together in an organization different from the same genes in *Acremonium chrysogenum* and *Penicillium chrysogenum*', *Mol Microbiol*, Vol. 5, No. 5, pp. 1125-1133.

Coque, J.J.R., Enguita, F.J., Cardoza, R.E., Martin, J.F. and Liras, P. (1996) 'Characterization of the cefF gene of *Nocardia lactamdurans* encoding a 3'-methylcephem hydroxylase different from the 7-cephem hydroxylase', *Applied Microbiology and Biotechnology*, Vol. 44, No. 5, pp. 605-609.

Coque, J.J.R., Perezllarena, F.J., Enguita, F.J., Fuente, J.L., Martin, J.F. and Liras, P. (1995b) 'Characterization of the cmcH genes of *Nocardia lactamdurans* and *Streptomyces clavuligerus* encoding a functional 3'-hydroxymethylcephem O-carbamoyltransferase for cephamycin biosynthesis', *Gene*, Vol. 162, No. 1, pp. 21-27.

Cortes, J., Martin, J.F., Castro, J.M., Laiz, L. and Liras, P. (1987) 'Purification and characterization of a 2-oxoglutarate-linked ATP-independent deacetoxycephalosporin C synthase of *Streptomyces lactamdurans*', *J Gen Microbiol*, Vol. 133, No. 11, pp. 3165-3174.

Crawford, L., Stepan, A.M., McAda, P.C., Rambosek, J.A., Conder, M.J., Vinci, V.A. and Reeves, C.D. (1995) 'Production of cephalosporin intermediates by feeding adipic acid to recombinant *Penicillium chrysogenum* strains expressing ring expansion activity', *Biotechnology (N Y)*, Vol. 13, No. 1, pp. 58-62.

Davies, J., Spiegelman, G.B. and Yim, G. (2006) 'The world of subinhibitory antibiotic concentrations', *Curr Opin Microbiol*, Vol. 9, No. 5, pp. 445-453.

Demain, A.L. (1973) 'Mutation and the production of secondary metabolites', *Adv Appl Microbiol*, Vol. 16, pp. 177-202.

Demain, A.L. (1983) 'New applications of microbial products', *Science*, Vol. 219, No. 4585, pp. 709-714.

Demain, A.L. and Elander, R.P. (1999) 'The beta-lactam antibiotics: past, present, and future', *Antonie Van Leeuwenhoek*, Vol. 75, No. 1-2, pp. 5-19.

Demirev, A.V., Lee, C.H., Jaishy, B.P., Nam, D.H. and Ryu, D.D. (2006) 'Substrate specificity of nonribosomal peptide synthetase modules responsible for the biosynthesis of the oligopeptide moiety of cephabacin in *Lysobacter lactamgenus*', *FEMS Microbiol Lett*, Vol. 255, No. 1, pp. 121-128.

Deresinski, S.C. (2008) 'The efficacy and safety of ceftobiprole in the treatment of complicated skin and skin structure infections: evidence from 2 clinical trials', *Diagn Microbiol Infect Dis*, Vol. 61, No. 1, pp. 103-109.

Diez, B., Barredo, J.L., Alvarez, E., Cantoral, J.M., van Solingen, P., Groenen, M.A., Veenstra, A.E. and Martin, J.F. (1989) 'Two genes involved in penicillin biosynthesis are linked in a 5.1 kb *Sal*I fragment in the genome of *Penicillium chrysogenum*', *Mol Gen Genet*, Vol. 218, No. 3, pp. 572-576.

Diez, B., Gutierrez, S., Barredo, J.L., van Solingen, P., van der Voort, L.H. and Martin, J.F. (1990) 'The cluster of penicillin biosynthetic genes. Identification and characterization of the *pcb*AB gene encoding the alpha-aminoadipyl-cysteinyl-valine synthetase and linkage to the *pcb*C and *pen*DE genes', *J Biol Chem*, Vol. 265, No. 27, pp. 16358-16365.

Dotzlaf, J.E. and Yeh, W.K. (1989) 'Purification and properties of deacetoxycephalosporin C synthase from recombinant *Escherichia coli* and its comparison with the native enzyme purified from *Streptomyces clavuligerus*', *J Biol Chem*, Vol. 264, No. 17, pp. 10219-10227.

Dubus, A., Llyod, M.D., Lee, H.J., Schofield, C.J., Baldwin, J.E. and Frere, J.M. (2001) 'Probing the penicillin sidechain selectivity of recombinant deacetoxycephalosporin C synthase', *Cellular and Molecular Life Sciences*, Vol. 58, No. 5-6, pp. 835-843.

Elander, R.P. (1983) *Strain improvement and preservation of beta-lactam producing microorganisms.* Springer-Verlag, New York.

Elander, R.P. (2003) 'Industrial production of beta-lactam antibiotics', *Applied Microbiology and Biotechnology*, Vol. 61, No. 5-6, pp. 385-392.

Enguita, F.J., Liras, P., Leitao, A.L. and Martin, J.F. (1996) 'Interaction of the two proteins of the methoxylation system involved in cephamycin C biosynthesis. Immunoaffinity, protein cross-linking, and fluorescence spectroscopy studies', *J Biol Chem*, Vol. 271, No. 52, pp. 33225-33230.

Evers, M.E., Trip, H., van den Berg, M.A., Bovenberg, R.A. and Driessen, A.J. (2004) 'Compartmentalization and transport in beta-lactam antibiotics biosynthesis', *Adv Biochem Eng Biotechnol*, Vol. 88, pp. 111-135.

Felnagle, E.A., Jackson, E.E., Chan, Y.A., Podevels, A.M., Berti, A.D., McMahon, M.D. and Thomas, M.G. (2008) 'Nonribosomal peptide synthetases involved in the production of medically relevant natural products', *Molecular Pharmaceutics*, Vol. 5, No. 2, pp. 191-211.

Fernandez, F.J., Cardoza, R.E., Montenegro, E., Velasco, J., Gutierrez, S. and Martin, J.F. (2003) 'The isopenicillin N acyltransferases of *Aspergillus nidulans* and *Penicillium chrysogenum* differ in their ability to maintain the 40-kDa alphabeta heterodimer in an undissociated form', *Eur J Biochem*, Vol. 270, No. 9, pp. 1958-1968.

Fernandez, M.J., Adrio, J.L., Piret, J.M., Wolfe, S., Ro, S. and Demain, A.L. (1999) 'Stimulatory effect of growth in the presence of alcohols on biotransformation of penicillin G into cephalosporin-type antibiotics by resting cells of *Streptomyces clavuligerus* NP1', *Applied Microbiology and Biotechnology*, Vol. 52, No. 4, pp. 484-488.

Fierro, F., Barredo, J.L., Diez, B., Gutierrez, S., Fernandez, F.J. and Martin, J.F. (1995) 'The penicillin gene cluster is amplified in tandem repeats linked by conserved hexanucleotide sequences', *Proc Natl Acad Sci U S A*, Vol. 92, No. 13, pp. 6200-6204.

Fleming, A. (2001) 'On the antibacterial action of cultures of a penicillium, with special reference to their use in the isolation of *B. influenzae*. 1929', *Bull World Health Organ*, Vol. 79, No. 8, pp. 780-790.

Gannon, J.C. (2000) 'The global infectious disease threat and its implications for the United States'. Available from: http://www.dni.gov /nic/special_globalinfectious.html

Gao, Q., Piret, J.M., Adrio, J.L. and Demain, A.L. (2003) 'Performance of a recombinant strain of *Streptomyces lividans* for bioconversion of penicillin G to deacetoxycephalosporin G', *Journal of Industrial Microbiology & Biotechnology*, Vol. 30, No. 3, pp. 190-194.

Giske, C.G., Ge, J. and Nordmann, P. (2009) 'Activity of cephalosporin CXA-101 (FR264205) and comparators against extended-spectrum-beta-lactamase-producing *Pseudomonas aeruginosa*', *J Antimicrob Chemother*, Vol. 64, No. 2, pp. 430-431.

Goo, K.S., Chua, C.S. and Sim, T.S. (2008a) 'A complete library of amino acid alterations at R306 in *Streptomyces clavuligerus* deacetoxycephalosporin C synthase demonstrates its structural role in the ring-expansion activity', *Proteins-Structure Function and Bioinformatics*, Vol. 70, No. 3, pp. 739-747.

Goo, K.S., Chua, C.S. and Sim, T.S. (2008b) 'Relevant double mutations in bioengineered *Streptomyces clavuligerus* deacetoxycephalosporin C synthase result in higher binding specificities which improve penicillin bioconversion', *Applied and Environmental Microbiology*, Vol. 74, No. 4, pp. 1167-1175.

Goo, K.S., Chua, C.S. and Sim, T.S. (2009) 'Directed evolution and rational approaches to improving *Streptomyces clavuligerus* deacetoxycephalosporin C synthase for cephalosporin production', *J Ind Microbiol Biotechnol*, Vol. 36, No. 5, pp. 619-633.

Goo, K.S. and Sim, T.S. (2011) 'Designing new beta-lactams: implications from their targets, resistance factors and synthesizing enzymes', *Curr Comput Aided Drug Des*, Vol. 7, No. 1, pp. 53-80.

Gouka, R.J., van Hartingsveldt, W., Bovenberg, R.A., van den Hondel, C.A. and van Gorcom, R.F. (1991) 'Cloning of the nitrate-nitrite reductase gene cluster of *Penicillium chrysogenum* and use of the *nia*D gene as a homologous selection marker', *J Biotechnol*, Vol. 20, No. 2, pp. 189-199.

Gould, I.M. and MacKenzie, F.M. (2002) 'Antibiotic exposure as a risk factor for emergence of resistance: the influence of concentration', *J Appl Microbiol*, Vol. 92 Suppl, pp. 78S-84S.

Guillen Schlippe, Y.V. and Hedstrom, L. (2005) 'A twisted base? The role of arginine in enzyme-catalyzed proton abstractions', *Arch Biochem Biophys*, Vol. 433, No. 1, pp. 266-278.

Gutierrez, S., Diez, B., Montenegro, E. and Martin, J.F. (1991) 'Characterization of the *Cephalosporium acremonium pcb*AB gene encoding alpha-aminoadipyl-cysteinyl-valine synthetase, a large multidomain peptide synthetase: linkage to the *pcb*C gene as a cluster of early cephalosporin biosynthetic genes and evidence of multiple functional domains', *J Bacteriol*, Vol. 173, No. 7, pp. 2354-2365.

Gutierrez, S., Velasco, J., Fernandez, F.J. and Martin, J.F. (1992) 'The *cef*G gene of *Cephalosporium acremonium* is linked to the *cef*EF gene and encodes a deacetylcephalosporin C acetyltransferase closely related to homoserine O-acetyltransferase', *J Bacteriol*, Vol. 174, No. 9, pp. 3056-3064.

Hamilton-Miller, J.M. (2000) 'Sir Edward Abraham's contribution to the development of the cephalosporins: a reassessment', *Int J Antimicrob Agents*, Vol. 15, No. 3, pp. 179-184.

Hamilton-Miller, J.M. (2008) 'Development of the semi-synthetic penicillins and cephalosporins', *Int J Antimicrob Agents*, Vol. 31, No. 3, pp. 189-192.

Hansch, C., Sammes, P.G. and Taylor, J.B. (1990) *Comprehensive medicinal chemistry : the rational design, mechanistic study & therapeutic application of chemical compounds*. Pergamon Press, Oxford ; New York.

Hausinger, R.P. (2004) 'Fe(II)/alpha-ketoglutarate-dependent hydroxylases and related enzymes', *Critical Reviews in Biochemistry and Molecular Biology*, Vol. 39, No. 1, pp. 21-68.

Henderson-Begg, S.K., Livermore, D.M. and Hall, L.M. (2006) 'Effect of subinhibitory concentrations of antibiotics on mutation frequency in *Streptococcus pneumoniae*', *J Antimicrob Chemother*, Vol. 57, No. 5, pp. 849-854.

Howard-Jones, A.R., Rutledge, P.J., Clifton, I.J., Adlington, R.M. and Baldwin, J.E. (2005) 'Unique binding of a non-natural L,L,L-substrate by isopenicillin N synthase', *Biochem Biophys Res Commun*, Vol. 336, No. 2, pp. 702-708.

Hsu, J.S., Yang, Y.B., Deng, C.H., Wei, C.L., Liaw, S.H. and Tsai, Y.C. (2004) 'Family shuffling of expandase genes to enhance substrate specificity for penicillin G', *Applied and Environmental Microbiology*, Vol. 70, No. 10, pp. 6257-6263.

Isogai, T., Fukagawa, M., Aramori, I., Iwami, M., Kojo, H., Ono, T., Ueda, Y., Kohsaka, M. and Imanaka, H. (1991) 'Construction of a 7-aminocephalosporanic acid (7ACA) biosynthetic operon and direct production of 7ACA in *Acremonium chrysogenum*', *Biotechnology (N Y)*, Vol. 9, No. 2, pp. 188-191.

Jensen, S.E., Westlake, D.W. and Wolfe, S. (1983) 'Analysis of penicillin N ring expansion activity from *Streptomyces clavuligerus* by ion-pair high-pressure liquid chromatography', *Antimicrob Agents Chemother*, Vol. 24, No. 3, pp. 307-312.

Jensen, S.E., Westlake, D.W. and Wolfe, S. (1985) 'Deacetoxycephalosporin C synthetase and deacetoxycephalosporin C hydroxylase are two separate enzymes in *Streptomyces clavuligerus*', *J Antibiot (Tokyo)*, Vol. 38, No. 2, pp. 263-265.

Kennedy, J. and Turner, G. (1996) 'delta-(L-alpha-aminoadipyl)-L-cysteinyl-D-valine synthetase is a rate limiting enzyme for penicillin production in *Aspergillus nidulans*', *Mol Gen Genet*, Vol. 253, No. 1-2, pp. 189-197.

Khosla, C. (2009) 'Structures and mechanisms of polyketide synthases', *J Org Chem*, Vol. 74, No. 17, pp. 6416-6420.

Kirby, W.M. (1944) 'Extraction of a highly potent penicillin inactivator from penicillin resistant staphylococci', *Science*, Vol. 99, No. 2579, pp. 452-453.

Kitano, K., Kintaka, K., Suzuki, S., Katamoto, K., Nara, K. and Nakao, Y. (1975) 'Studies on production of beta-lactam antibiotics .1. Screening of microorganisms capable of producing beta-lactam antibiotics', *Journal of Fermentation Technology*, Vol. 53, No. 6, pp. 327-338.

Laudano, J.B. (2011) 'Ceftaroline fosamil: a new broad-spectrum cephalosporin', *J Antimicrob Chemother*, Vol. 66 Suppl 3, pp. iii11-18.

Lee, H.J., Lloyd, M.D., Harlos, K., Clifton, I.J., Baldwin, J.E. and Schofield, C.J. (2001) 'Kinetic and crystallographic studies on deacetoxycephalosporin C synthase (DAOCS)', *Journal of Molecular Biology*, Vol. 308, No. 5, pp. 937-948.

Lee, H.J., Schofield, C.J. and Lloyd, M.D. (2002) 'Active site mutations of recombinant deacetoxycephalosporin C synthase', *Biochemical and Biophysical Research Communications*, Vol. 292, No. 1, pp. 66-70.

Lein, J. (1986) *The Panlabs Penicillium strain improvement program.* Butterworths, Stoneham, MA.

Liras, P. (1999) 'Biosynthesis and molecular genetics of cephamycins - Cephamycins produced by Actinomycetes', *Antonie Van Leeuwenhoek International Journal of General and Molecular Microbiology*, Vol. 75, No. 1-2, pp. 109-124.

Liras, P. and Demain, A.L. (2009) 'Chapter 16. Enzymology of beta-lactam compounds with cephem structure produced by actinomycete', *Methods Enzymol*, Vol. 458, pp. 401-429.

Liu, Y., Gong, G., Zhu, C., Zhu, B. and Hu, Y. (2010) 'Environmentally safe production of 7-ACA by recombinant *Acremonium chrysogenum*', *Curr Microbiol*, Vol. 61, No. 6, pp. 609-614.

Livermore, D.M., Mushtaq, S., Ge, Y. and Warner, M. (2009) 'Activity of cephalosporin CXA-101 (FR264205) against *Pseudomonas aeruginosa* and *Burkholderia cepacia* group strains and isolates', *Int J Antimicrob Agents*, Vol. 34, No. 5, pp. 402-406.

Llarrull, L.I., Fisher, J.F. and Mobashery, S. (2009) 'Molecular basis and phenotype of methicillin resistance in *Staphylococcus aureus* and insights into new beta-lactams that meet the challenge', *Antimicrob Agents Chemother*, Vol. 53, No. 10, pp. 4051-4063.

Lloyd, M.D., Lee, H.J., Harlos, K., Zhang, Z.H., Baldwin, J.E., Schofield, C.J., Charnock, J.M., Garner, C.D., Hara, T., Terwisscha van Scheltinga, A.C., Valegard, K., Viklund, J.A., Hajdu, J., Andersson, I., Danielsson, A. and Bhikhabhai, R. (1999) 'Studies on the active site of deacetoxycephalosporin C synthase', *J Mol Biol*, Vol. 287, No. 5, pp. 943-960.

Loder, P.B. and Abraham, E.P. (1971) 'Biosynthesis of peptides containing -aminoadipic acid and cysteine in extracts of a Cephalosporium sp', *Biochem J*, Vol. 123, No. 3, pp. 477-482.

Loke, P., Sim, J. and Sim, T.S. (1997) 'Functional analysis of a conserved aspartate D218 in *Cephalosporium acremonium* isopenicillin N synthase', *FEMS Microbiol Lett*, Vol. 157, No. 1, pp. 137-140.

Luengo, J.M. (1995) 'Enzymatic synthesis of hydrophobic penicillins', *J Antibiot (Tokyo)*, Vol. 48, No. 11, pp. 1195-1212.

Maeda, K., Luengo, J.M., Ferrero, O., Wolfe, S., Lebedev, M.Y., Fang, A. and Demain, A.L. (1995) 'The substrate-specificity of deacetoxycephalosporin C synthase (expandase) of *Streptomyces clavuligerus* is extremely narrow', *Enzyme and Microbial Technology*, Vol. 17, No. 3, pp. 231-234.

Marshall, W.F. and Blair, J.E. (1999) 'The cephalosporins', *Mayo Clin Proc*, Vol. 74, No. 2, pp. 187-195.

Martin, J.F. (1998) 'New aspects of genes and enzymes for beta-lactam antibiotic biosynthesis', *Applied Microbiology and Biotechnology*, Vol. 50, No. 1, pp. 1-15.

Martin, J.F. and Gutierrez, S. (1995) 'Genes for beta-lactam antibiotic biosynthesis', *Antonie Van Leeuwenhoek International Journal of General and Molecular Microbiology*, Vol. 67, No. 2, pp. 181-200.

Martin, J.F., Ullan, R.V. and Casqueiro, J. (2004) 'Novel genes involved in cephalosporin biosynthesis: The three-component isopenicillin N epimerase system': *Molecular Biotechnology of Fungal Beta-Lactam Antibiotics and Related Peptide Synthetases*, pp. 91-109.

Martin, J.F., Ullan, R.V. and Garcia-Estrada, C. (2010) 'Regulation and compartmentalization of beta-lactam biosynthesis', *Microb Biotechnol*, Vol. 3, No. 3, pp. 285-299.

Matsuhashi, M., Song, M.D., Ishino, F., Wachi, M., Doi, M., Inoue, M., Ubukata, K., Yamashita, N. and Konno, M. (1986) 'Molecular cloning of the gene of a penicillin-binding protein supposed to cause high resistance to beta-lactam antibiotics in *Staphylococcus aureus*', *J Bacteriol*, Vol. 167, No. 3, pp. 975-980.

Menzella, H.G. and Reeves, C.D. (2007) 'Combinatorial biosynthesis for drug development', *Curr Opin Microbiol*, Vol. 10, No. 3, pp. 238-245.

Mesaros, N., Nordmann, P., Plesiat, P., Roussel-Delvallez, M., Van Eldere, J., Glupczynski, Y., Van Laethem, Y., Jacobs, F., Lebecque, P., Malfroot, A., Tulkens, P.M. and Van Bambeke, F. (2007) '*Pseudomonas aeruginosa*: resistance and therapeutic options at the turn of the new millennium', *Clin Microbiol Infect*, Vol. 13, No. 6, pp. 560-578.

Mildvan, A.S., Weber, D.J. and Kuliopulos, A. (1992) 'Quantitative interpretations of double mutations of enzymes', *Arch Biochem Biophys*, Vol. 294, No. 2, pp. 327-340.

Miller, L.A., Ratnam, K. and Payne, D.J. (2001) 'Beta-lactamase-inhibitor combinations in the 21st century: current agents and new developments', *Curr Opin Pharmacol*, Vol. 1, No. 5, pp. 451-458.

Minambres, B., Martinez-Blanco, H., Olivera, E.R., Garcia, B., Diez, B., Barredo, J.L., Moreno, M.A., Schleissner, C., Salto, F. and Luengo, J.M. (1996) 'Molecular cloning and expression in different microbes of the DNA encoding *Pseudomonas putida* U phenylacetyl-CoA ligase. Use of this gene to improve the rate of benzylpenicillin biosynthesis in *Penicillium chrysogenum*', *J Biol Chem*, Vol. 271, No. 52, pp. 33531-33538.

Morin, R.B., Roeske, R.W., Flynn, E.H. and Jackson, B.G. (1962) 'Chemistry of cephalosporin antibiotics .1. 7-Aminocephalosporanic acid from cephalosporin C', *Journal of the American Chemical Society*, Vol. 84, No. 17, pp. 3400-3401.

Murano, K., Yamanaka, T., Toda, A., Ohki, H., Okuda, S., Kawabata, K., Hatano, K., Takeda, S., Akamatsu, H., Itoh, K., Misumi, K., Inoue, S. and Takagi, T. (2008) 'Structural requirements for the stability of novel cephalosporins to AmpC beta-lactamase based on 3D-structure', *Bioorg Med Chem*, Vol. 16, No. 5, pp. 2261-2275.

Nagarajan, R., Boeck, L.D., Gorman, M., Hamill, R.L., Higgens, C.E., Hoehn, M.M., Stark, W.M. and Whitney, J.G. (1971) 'Beta-lactam antibiotics from *Streptomyces*', *J Am Chem Soc*, Vol. 93, No. 9, pp. 2308-2310.

Nayler, J.H. (1991) 'Semi-synthetic approaches to novel penicillins', *Trends Biochem Sci*, Vol. 16, No. 6, pp. 234-237.

Newbert, R.W., Barton, B., Greaves, P., Harper, J. and Turner, G. (1997) 'Analysis of a commercially improved *Penicillium chrysogenum* strain series: involvement of recombinogenic regions in amplification and deletion of the penicillin biosynthesis gene cluster', *J Ind Microbiol Biotechnol*, Vol. 19, No. 1, pp. 18-27.

Nijland, J.G., Ebbendorf, B., Woszczynska, M., Boer, R., Bovenberg, R.A. and Driessen, A.J. (2010) 'Nonlinear biosynthetic gene cluster dose effect on penicillin production by *Penicillium chrysogenum*', *Appl Environ Microbiol*, Vol. 76, No. 21, pp. 7109-7115.

Nozaki, Y., Katayama, N., Tsubotani, S., Ono, H. and Okazaki, H. (1985) 'Cephabacin M1-6, new 7-methoxycephem antibiotics of bacterial origin.

I. A producing organism, fermentation, biological activities, and mode of action', *J Antibiot (Tokyo)*, Vol. 38, No. 9, pp. 1141-1151.

Nozaki, Y., Okonogi, K., Katayama, N., Ono, H., Harada, S., Kondo, M. and Okazaki, H. (1984) 'Cephabacins, new cephem antibiotics of bacterial origin. IV. Antibacterial activities, stability to beta-lactamases and mode of action', *J Antibiot (Tokyo)*, Vol. 37, No. 12, pp. 1555-1565.

O'Sullivan, J., Aplin, R.T., Stevens, C.M. and Abraham, E.P. (1979) 'Biosynthesis of a 7-alpha-methoxycephalosporin. Incorporation of molecular oxygen', *Biochem J*, Vol. 179, No. 1, pp. 47-52.

Oh, B., Kim, M., Yoon, J., Chung, K., Shin, Y., Lee, D. and Kim, Y. (2003) 'Deacylation activity of cephalosporin acylase to cephalosporin C is improved by changing the side-chain conformations of active-site residues', *Biochem Biophys Res Commun*, Vol. 310, No. 1, pp. 19-27.

Ono, H., Nozaki, Y., Katayama, N. and Okazaki, H. (1984) 'Cephabacins, new cephem antibiotics of bacterial origin. I. Discovery and taxonomy of the producing organisms and fermentation', *J Antibiot (Tokyo)*, Vol. 37, No. 12, pp. 1528-1535.

Oster, L.M., van Scheltinga, A.C., Valegard, K., Hose, A.M., Dubus, A., Hajdu, J. and Andersson, I. (2004) 'Conformational flexibility of the C terminus with implications for substrate binding and catalysis revealed in a new crystal form of deacetoxycephalosporin C synthase', *J Mol Biol*, Vol. 343, No. 1, pp. 157-171.

Otten, L.G., Sio, C.F., Vrielink, J., Cool, R.H. and Quax, W.J. (2002) 'Altering the substrate specificity of cephalosporin acylase by directed evolution of the Beta -subunit', *J Biol Chem*, Vol. 277, No. 44, pp. 42121-42127.

Palissa, H., von Dohren, H., Kleinkauf, H., Ting, H.H. and Baldwin, J.E. (1989) 'Beta-lactam biosynthesis in a gram-negative eubacterium: purification and characterization of isopenicillin N synthase from *Flavobacterium* sp. strain SC 12.154', *J Bacteriol*, Vol. 171, No. 10, pp. 5720-5728.

Penalva, M.A., Rowlands, R.T. and Turner, G. (1998) 'The optimization of penicillin biosynthesis in fungi', *Trends Biotechnol*, Vol. 16, No. 11, pp. 483-489.

Pollegioni, L., Lorenzi, S., Rosini, E., Marcone, G.L., Molla, G., Verga, R., Cabri, W. and Pilone, M.S. (2005) 'Evolution of an acylase active on cephalosporin C', *Protein Sci*, Vol. 14, No. 12, pp. 3064-3076.

Queener, S.W. (1990) 'Molecular biology of penicillin and cephalosporin biosynthesis', *Antimicrob Agents Chemother*, Vol. 34, No. 6, pp. 943-948.

Raper, K.B., Alexander, D.F. and Coghill, R.D. (1944) 'Penicillin: II. Natural variation and penicillin production in *Penicillium notatum* and allied species', *J Bacteriol*, Vol. 48, No. 6, pp. 639-659.

Roach, P.L., Clifton, I.J., Fulop, V., Harlos, K., Barton, G.J., Hajdu, J., Andersson, I., Schofield, C.J. and Baldwin, J.E. (1995) 'Crystal structure of isopenicillin N synthase is the first from a new structural family of enzymes', *Nature*, Vol. 375, No. 6533, pp. 700-704.

Roach, P.L., Clifton, I.J., Hensgens, C.M., Shibata, N., Schofield, C.J., Hajdu, J. and Baldwin, J.E. (1997) 'Structure of isopenicillin N synthase complexed with substrate and the mechanism of penicillin formation', *Nature*, Vol. 387, No. 6635, pp. 827-30.

Rolinson, G.N. (1998) 'Forty years of beta-lactam research', *J Antimicrob Chemother*, Vol. 41, No. 6, pp. 589-603.

Rolinson, G.N. and Geddes, A.M. (2007) 'The 50th anniversary of the discovery of 6-aminopenicillanic acid (6-APA)', *Int J Antimicrob Agents*, Vol. 29, No. 1, pp. 3-8.

Schmidt, F.-R. (2010) *The beta-lactam antibiotics: current situation and future prospects in manufacture and therapy*. Springer-Verlag Heidelberg.

Sheehan, J.C. and Henry-Logan, K.R. (1957) 'The total synthesis of penicillin-V', *Journal of the American Chemical Society*, Vol. 79, No. 5, pp. 1262-1263.

Shen, Y.Q., Wolfe, S. and Demain, A.L. (1984) 'Deacetoxycephalosporin-C synthetase - importance of order of cofactor/reactant addition', *Enzyme and Microbial Technology*, Vol. 6, No. 9, pp. 402-404.

Sim, J. and Sim, T.S. (2001) 'In vitro conversion of penicillin G and ampicillin by recombinant *Streptomyces clavuligerus* NRRL 3585 deacetoxycephalosporin C synthase', *Enzyme and Microbial Technology*, Vol. 29, No. 4-5, pp. 240-245.

Sim, J., Wong, E., Chin, H.S. and Sim, T.S. (2003) 'Conserved structural modules and bonding networks in isopenicillin N synthase related non-haem iron-dependent oxygenases and oxidases', *Journal of Molecular Catalysis B-Enzymatic*, Vol. 23, No. 1, pp. 17-27.

Skatrud, P.L., Queener, S.W., Fisher, D.L. and Chapman, J.L. (1987) 'Strain improvement studies in *Penicillium chrysogenum* using the cloned *P. chrysogenum* isopenicillin N synthase gene and the AmdS gene of *Aspergillus nidulans*', *Society for Industrial Microbiology News*, Vol. 37, p. 77.

Sonawane, V.C. (2006) 'Enzymatic modifications of cephalosporins by cephalosporin acylase and other enzymes', *Crit Rev Biotechnol*, Vol. 26, No. 2, pp. 95-120.

Stapley, E.O., Jackson, M., Hernandez, S., Zimmerman, S.B., Currie, S.A., Mochales, S., Mata, J.M., Woodruff, H.B. and Hendlin, D. (1972) 'Cephamycins, a new family of beta-lactam antibiotics. I. Production by actinomycetes, including *Streptomyces lactamdurans* sp. n', *Antimicrob Agents Chemother*, Vol. 2, No. 3, pp. 122-131.

Stein, R.A., Goetz, R.M. and Ganea, G.M. (2009) 'Ceftobiprole: a new beta-lactam antibiotic', *Int J Clin Pract*, Vol. 63, No. 6, pp. 930-943.

Strateva, T. and Yordanov, D. (2009) '*Pseudomonas aeruginosa* - a phenomenon of bacterial resistance', *J Med Microbiol*, Vol. 58, No. Pt 9, pp. 1133-1148.

Strieker, M., Tanovic, A. and Marahiel, M.A. (2010) 'Nonribosomal peptide synthetases: structures and dynamics', *Curr Opin Struct Biol*, Vol. 20, No. 2, pp. 234-240.

Tan, D.S. and Sim, T.S. (1996) 'Functional analysis of conserved histidine residues in *Cephalosporium acremonium* isopenicillin N synthase by site-directed mutagenesis', *J Biol Chem*, Vol. 271, No. 2, pp. 889-894.

Tan, J.S. and File, T.M., Jr. (1995) 'Antipseudomonal penicillins', *Med Clin North Am*, Vol. 79, No. 4, pp. 679-693.

Theilgaard, H., van Den Berg, M., Mulder, C., Bovenberg, R. and Nielsen, J. (2001) 'Quantitative analysis of *Penicillium chrysogenum* Wis54-1255 transformants overexpressing the penicillin biosynthetic genes', *Biotechnol Bioeng*, Vol. 72, No. 4, pp. 379-388.

Titelman, E., Karlsson, I.M., Ge, Y. and Giske, C.G. (2011) '*In vitro* activity of CXA-101 plus tazobactam (CXA-201) against CTX-M-14- and CTX-M-15-producing *Escherichia coli* and *Klebsiella pneumoniae*', *Diagn Microbiol Infect Dis*, Vol. 70, No. 1, pp. 137-141.

Tobin, M.B., Kovacevic, S., Madduri, K., Hoskins, J.A., Skatrud, P.L., Vining, L.C., Stuttard, C. and Miller, J.R. (1991) 'Localization of the lysine epsilon-aminotransferase (*lat*) and delta-(L-alpha-aminoadipyl)-L-cysteinyl-D-valine synthetase (*pcb*AB) genes from *Streptomyces clavuligerus* and production of lysine epsilon-aminotransferase activity in *Escherichia coli*', *J Bacteriol*, Vol. 173, No. 19, pp. 6223-6229.

Toda, A., Ohki, H., Yamanaka, T., Murano, K., Okuda, S., Kawabata, K., Hatano, K., Matsuda, K., Misumi, K., Itoh, K., Satoh, K. and Inoue, S. (2008) 'Synthesis and SAR of novel parenteral anti-pseudomonal cephalosporins: discovery of FR264205', *Bioorg Med Chem Lett*, Vol. 18, No. 17, pp. 4849-4852.

Ubukata, K., Nonoguchi, R., Matsuhashi, M. and Konno, M. (1989) 'Expression and inducibility in *Staphylococcus aureus* of the *mec*A gene, which encodes a methicillin-resistant *S. aureus*-specific penicillin-binding protein', *J Bacteriol*, Vol. 171, No. 5, pp. 2882-2885.

Ubukata, K., Yamashita, N. and Konno, M. (1985) 'Occurrence of a beta-lactam-inducible penicillin-binding protein in methicillin-resistant staphylococci', *Antimicrob Agents Chemother*, Vol. 27, No. 5, pp. 851-857.

Ullan, R.V., Campoy, S., Casqueiro, J., Fernandez, F.J. and Martin, J.F. (2007) 'Deacetylcephalosporin C production in *Penicillium chrysogenum* by expression of the isopenicillin N epimerization, ring expansion, and acetylation genes', *Chemistry & Biology*, Vol. 14, No. 3, pp. 329-339.

Ullan, R.V., Casqueiro, J., Banuelos, O., Fernandez, F.J., Gutierrez, S. and Martin, J.F. (2002) 'A novel epimerization system in fungal secondary metabolism involved in the conversion of isopenicillin N into penicillin N in *Acremonium chrysogenum*', *Journal of Biological Chemistry*, Vol. 277, No. 48, pp. 46216-46225.

Usher, J.J., Lewis, M.A., Hughes, D.W. and Compton, B.J. (1988) 'Development of the cephalosporin-C fermentation taking into account the instability of cephalosporin C', *Biotechnology Letters*, Vol. 10, No. 8, pp. 543-548.

Usui, S. and Yu, C.A. (1989) 'Purification and properties of isopenicillin N epimerase from *Streptomyces clavuligerus*', *Biochim Biophys Acta*, Vol. 999, No. 1, pp. 78-85.

Valegard, K., Terwisscha van Scheltinga, A.C., Dubus, A., Ranghino, G., Oster, L.M., Hajdu, J. and Andersson, I. (2004) 'The structural basis of cephalosporin formation in a mononuclear ferrous enzyme', *Nat Struct Mol Biol*, Vol. 11, No. 1, pp. 95-101.

Valegard, K., van Scheltinga, A.C.T., Lloyd, M.D., Hara, T., Ramaswamy, S., Perrakis, A., Thompson, A., Lee, H.J., Baldwin, J.E., Schofield, C.J., Hajdu, J. and Andersson, I. (1998) 'Structure of a cephalosporin synthase', *Nature*, Vol. 394, No. 6695, pp. 805-809.

van de Kamp, M., Driessen, A.J.M. and Konings, W.N. (1999) 'Compartmentalization and transport in beta-lactam antibiotic biosynthesis by filamentous fungi', *Antonie Van Leeuwenhoek International Journal of General and Molecular Microbiology*, Vol. 75, No. 1-2, pp. 41-78.

Veenstra, A.E., van Solingen, P., Bovenberg, R.A. and van der Voort, L.H. (1991) 'Strain improvement of *Penicillium chrysogenum* by recombinant DNA techniques', *J Biotechnol*, Vol. 17, No. 1, pp. 81-90.

Velasco, J., Adrio, J.L., Moreno, M.A., Diez, B., Soler, G. and Barredo, J.L. (2000) 'Environmentally safe production of 7-aminodeacetoxy-cephalosporanic acid (7-ADCA) using recombinant strains of *Acremonium chrysogenum*', *Nature Biotechnology*, Vol. 18, No. 8, pp. 857-861.

Villegas-Estrada, A., Lee, M., Hesek, D., Vakulenko, S.B. and Mobashery, S. (2008) 'Co-opting the cell wall in fighting methicillin-resistant *Staphylococcus aureus*: potent inhibition of PBP 2a by two anti-MRSA beta-lactam antibiotics', *J Am Chem Soc*, Vol. 130, No. 29, pp. 9212-9213.

Walkty, A., Decorby, M., Nichol, K., Karlowsky, J.A., Hoban, D.J. and Zhanel, G.G. (2008) 'In vitro activity of ceftobiprole against clinical isolates of *Pseudomonas aeruginosa* obtained from Canadian intensive care unit (ICU) patients as part of the CAN-ICU Study', *J Antimicrob Chemother*, Vol. 62, No. 1, pp. 206-208.

Wei, C.L., Yang, Y.B., Deng, C.H., Liu, W.C., Hsu, J.S., Lin, Y.C., Liaw, S.H. and Tsai, Y.C. (2005) 'Directed evolution of *Streptomyces clavuligerus* deacetoxycephalosporin C synthase for enhancement of penicillin G expansion', *Applied and Environmental Microbiology*, Vol. 71, No. 12, pp. 8873-8880.

Wei, C.L., Yang, Y.B., Wang, W.C., Liu, W.C., Hsu, J.S. and Tsai, Y.C. (2003) 'Engineering *Streptomyces clavuligerus* deacetoxycephalosporin C synthase for optimal ring expansion activity toward penicillin G', *Applied and Environmental Microbiology*, Vol. 69, No. 4, pp. 2306-2312.

Wells, J.A. (1990) 'Additivity of mutational effects in proteins', *Biochemistry*, Vol. 29, No. 37, pp. 8509-8517.

Whiteman, P.A., Abraham, E.P., Baldwin, J.E., Fleming, M.D., Schofield, C.J., Sutherland, J.D. and Willis, A.C. (1990) 'Acyl coenzyme A: 6-aminopenicillanic acid acyltransferase from *Penicillium chrysogenum* and *Aspergillus nidulans*', *FEBS Lett*, Vol. 262, No. 2, pp. 342-344.

Whitney, J.G., Brannon, D.R., Mabe, J.A. and Wicker, K.J. (1972) 'Incorporation of labeled precursors into A16886B, a novel -lactam antibiotic produced by *Streptomyces clavuligerus*', *Antimicrob Agents Chemother*, Vol. 1, No. 3, pp. 247-251.

Wolfe, S., Demain, A.L., Jensen, S.E. and Westlake, D.W. (1984) 'Enzymatic approach to syntheses of unnatural beta-lactams', *Science*, Vol. 226, No. 4681, pp. 1386-1392.

Wright, A.J. (1999) 'The penicillins', *Mayo Clin Proc*, Vol. 74, No. 3, pp. 290-307.

Wu, X.B., Fan, K.Q., Wang, Q.H. and Yang, K.Q. (2005) 'C-terminus mutations of *Acremonium chrysogenum* deacetoxy/deacetylcephalosporin C synthase with improved activity toward penicillin analogs', *Fems Microbiology Letters*, Vol. 246, No. 1, pp. 103-110.

Wu, X.B., Tian, X.Y., Ji, J.J., Wu, W.B., Fan, K.Q. and Yang, K.Q. (2011) 'Saturation mutagenesis of *Acremonium chrysogenum* deacetoxy/deacetylcephalosporin C synthase R308 site confirms its role in controlling substrate specificity', *Biotechnol Lett*, Vol. 33, No. 4, pp. 805-812.

Zamorano, L., Juan, C., Fernandez-Olmos, A., Ge, Y., Canton, R. and Oliver, A. (2010) 'Activity of the new cephalosporin CXA-101 (FR264205) against *Pseudomonas aeruginosa* isolates from chronically infected cystic fibrosis patients', *Clin Microbiol Infect.* Vol. 16, No. 9, pp.1482-1487.

Zhou, J., Kelly, W.L., Bachmann, B.O., Gunsior, M., Townsend, C.A. and Solomon, E.I. (2001) 'Spectroscopic studies of substrate interactions with clavaminate synthase 2, a multifunctional alpha-KG-dependent non-heme iron enzyme: correlation with mechanisms and reactivities', *J Am Chem Soc*, Vol. 123, No. 30, pp. 7388-7398.

In: Penicillin ISBN: 978-1-62100-233-8
Editors: G. Barresi and P. Komulainen

Chapter 2

DEVELOPMENTS IN THE SEPARATION AND RECOVERY OF PENICILLIN G USING HOLLOW-FIBER MEMBRANE CONTACTORS

Monwar Hossain*
Department of Chemical & Petroleum Engineering
United Arab Emirates University, Al Ain, United Arab Emirates

ABSTRACT

The separation and recovery of pharmaceuticals such as penicillin G from fermentation media is very important. For this purpose, the research for an efficient, cost-effective and environmentally-friendly process continued over the last two decades. Various solvents including kerosene, butyl acetate, tributyl phosphate (TBP), octanol and their mixtures have shown to have the potential but they have the disadvantages of being toxic and expensive. Recently, non-traditional solvents like vegetable oils have been examined as they can overcome some of the above-mentioned disadvantages. In this chapter, results of research are presented for various organic solvents including non-traditional ones like vegetable oils for the separation of penicillin G using physical and reactive extractions. In the first step of the investigation, the effects of organic solvents on apparent distribution of penicillin G either from a single-component solution or from a synthetic media are determined. Then the effects of the addition of an ionic "carrier", such as Amberlite-LA- in the solvent on the

* E-mail: mmonwar@uaeu.ac.ae.

enhancement of the distribution factor and on the degree of extraction are examined. The carrier-solvent systems have shown to have considerable improvement in the percentages of separation of penicillin G. As a demonstration of the improved process, the results are presented for Amberlite-LA2 in various organic solvents including sunflower oil phase. With the addition of a small amount (5%v/v) of the carrier in sunflower oil, the separation of penicillin G from the fermentation can be performed at the natural pH of fermentation (that is without adding any chemicals for pH adjustment). The percentage extraction of approx. 40-45% was achieved at a circulating flow rate of 13-16 L/h in a hollow-fibre membrane contactor. This result is similar to those reported by many investigators using toxic solvents (such as kerosene). The performance of the carrier-solvent system can be improved if pH and temperature are adjusted to lower values. This would require the media solution to be adjusted again for the fermentation of penicillin G as it requires a pH of 6.5. For recovery from the organic phase, aqueous solutions of sodium carbonate and sodium chloride have shown good results. An aqueous solution of sodium carbonate at pH 11 is able to recover significantly than sodium chloride solution and in a shorter process time of about 1.5 hours. From these studies it can be concluded that sunflower-based system is the preferred organic solvent for extraction because other solvents are very expensive (much more than that of sunflower oil) and do not have renewable sources. This new system offers many process advantages being less toxic, less corrosive, equally effective, environmentally-safe and operator-friendly.

NOTATION

A	surface area (m^2)
B	defined in Equation (12)
C	concentration of penicillin G (mmol/L)
di, do	inner diameter, outer diameter of a hollow fiber (cm)
D_E	distribution coefficient of penicillin G defined in Eq. (6)
D_P	diffusivity of penicillin G defined in Eq. (16)
D_X	diffusivity of penicillin G defined in Eq. (18)
E (%)	percent of extraction of Penicillin G, defined in Eq. (13)
k_f	mass transfer coefficient of penicillin G defined in Eq. (9)
k_{mf}	mass transfer coefficient of penicillin G defined in Eq. (10)
k_o	mass transfer coefficient of penicillin G defined in Eq. (11)
H+	hydrogen ion
HP	undissociated penicillin G

K_{eq}	equilibrium constant, defined in Eq. (4)
K_{of}	mass transfer coefficient in Equation (10), (cm/sec)
L	length of the fiber (cm)
LA	Amberlite LA-2 carrier
n_f	number of fibers
N	steady state flux, defined in Eqn. (7)
P	penicillin G anion
t	time (s)
q	flow rate, L/s
u	linear velocity in the hollow fibers (cm/s)
v	velocity of feed solution (cm/s)
v	velocity of feed solution (cm/s)
v	velocity of feed solution (cm/s)
v	velocity of feed solution (cm/s)
V	volume (L)
Z	axial distance in the module (cm)

Subscripts

aq	aqueous phase
A	Amine carrier
C	penicillin-amine complex
f	feed solution
fi	feed side at aqueous-organic interface
fo	organic side at aqueous-organic interface
f	final condition
HP	undissociated penicillin
i	initial condition
o	organic phase

Superscripts

m	module
t	tank

Greek Symbols

ε	membrane porosity
μ	viscosity of the solution
σ	membrane thickness
τ	tortuosity of membrane pores

INTRODUCTION

Penicillin is an extremely important antibiotic, and penicillin G (Pen G) is the raw material of the semi-synthetic penicillins. Over 20 countries manufacture over 11,000 tons of penicillin per year [1]. Its wide usage is a result of its lack of toxicity and irritancy [2]. Penicillin G is commonly produced by submerged aerobic fermentation using strains of *penicillium chryogenum*. As in most bioconversions, the production is inhibited by the presence of excess product, so for maximum production it is desirable to remove the product in situ – In Situ Product Recovery (ISPR) [3]. Penicillin G is a weak monocarboxylic acid (pKa = 2.75) [4]. It has traditionally been extracted from the fermentation broth by physical extraction with butyl acetate [4-6]. For physical extraction, penicillin G must be in the undissociated state, so the pH of the aqueous phase needs to be adjusted to around 2.0-2.5 [3]. At this low pH however, penicillin G is unstable, and decomposes irreversibly [7]. For this reason, the extraction is often carried out at low temperatures (0-5°C). At a pH of 2.0 and ambient temperature, 1% of penicillin G present decomposes in 10 s. [7]. Even at low temperature extraction, there is usually 10-15% overall loss of product. Other disadvantages of this system include high consumption of butyl acetate, high power consumption for recovery of butyl acetate from raffinate (due to the solubility of butyl acetate in aqueous phase) and a low recovery yield (a maximum of 85%) of penicillin G [5].

Various traditional methods for the extraction of penicillin G have been reported such as precipitation, ion exchange process, adsorption, diffusion dialysis, microcapsules, esterification as well as a simulated moving bed process [4, 6, 8-11]. These methods have several disadvantages including high cost, and they produce large volumes of waste, require multiple steps, difficult regeneration technique and many of them operate with low efficiency under practical conditions.

Extractions that combine reaction with separation (reactive liquid-liquid extraction, RLLE) have much greater selectivity and yield compared to other techniques [8-10]. The carrier molecule can selectively and reversibly react with the solute, increasing both the amount of extracts and the selectivity [9, 12-16]. It also enables penicillin G to be extracted at a higher pH, where this is more stable, so decomposition losses can be greatly reduced and thus may provide increased productivity.

Amine compounds have been found useful as extractants for the separation of antibiotics, e.g. penicillin G [9, 10, 17, 18]. They provide high efficiency and selectivity. Secondary, tertiary and quaternary amines and their mixtures have been employed for this purpose. An organic solvent is required to dissolve the reaction product, (also known as diluent in the literature when carrier is present in the system) and a modifier is required to control the viscosity and to stop formation of any third phase. Active polar and proton-donating diluents as alcohols have been shown to be the most suitable diluents for amines as they show high distribution coefficient. The reaction mechanism may change with the combination of the extractant and solvent type. A number of extractant-diluent combinations have been used in the literature. The solvents include hydrocarbons (methyl isobutylketone, hexane, toluene, kerosene), halogenated organics like dichloromethane, butyl acetate and tributyl phosphate and the experiments were conducted in small-scale experiments.

Many membrane-based systems using amines as extractants and kerosene and similar hydrocarbon as diluent have been reported for extraction [10 - 14]. These systems work well but have some disadvantages of solvent toxicity and corrosiveness, and swelling problems in membrane contactors.

The liquid membrane processes using hollow fiber module contactors (HFMCs) offer an advantage of large interfacial area per unitt volume and an option for continuous operation. Because of their attractiveness, HFMCs are being considered for the recovery and the extraction of organic solutes from dilute solutions [17 - 22].

As mentioned earlier, the RLLE method using microporous HFMCs may potentially overcome many of the disadvantages and provide a better alternative for the recovery of fermentation products [8, 11, 21 - 24]. In recent papers, processes based on RLLE in HFMM have been found to be competitive from the process, economic and environmental points of view [12, 25, 26]. The advantages of the membrane mass transfer process over the conventional systems are:

- Selectivity and flexibility of extraction
- *in situ* application to reduce any inhibitory effect
- Reduction of number of steps (improved productivity)
- Use of operator and environmentally-friendly organic system
- Minimal dispersion of phases (less contamination)
- Recycle of extracting media and generation of smaller wastes
- Lower temperature operation requiring less energy
- Ability to operate on identical density systems
- Availability of large-scale module (i.e. easy scale up methods).

The main aim of this chapter is to present (a) the investigations of the reactive extraction of penicillin G in comparison to physical extraction, (b) the effects of the operating conditions on extraction and re-extraction, and (c) performance studies in HFMCs in evaluating the process of penicillin G separation, especially with sunflower oil which has more potential than other solvents.

To understand and explore more of this process the following aims are also included:

- to identify and examine a less toxic, environmentally-friendly solvent and a carrier or a mixture of carriers for extraction of penicillin G
- to determine the effect of extraction variables (initial feed concentration, pH and organic phase composition) covering the range of real fermentation,
- to determine the kinetics and mechanism of reactive liquid-liquid extraction
- to develop a mathematical model for the mass transfer processes in a small pilot-scale contactor
- to evaluate the performance of the less toxic solvent for extraction under various conditions including fermentation conditions (i.e. in presence of biomedium and biomass).
- to determine the values of the mass transfer coefficient in the process
- to compare the results of the hollow-fibre experiments with other solvent systems.

The results show that the new extraction system has the potential to overcome some of the disadvantages mentioned above. More research is required to optimise the experimental conditions, to develop a more comprehensive mathematical model including extraction and re-extraction and obtain performance data with "real" (rather than synthetic data) system.

In the next section, mechanism and kinetics are presented for liquid-liquid extraction of penicillin G.

MECHANISM AND KINETICS OF REACTIVE EXTRACTION

Reactive liquid-liquid extraction is carried out using a carrier molecule which forms a complex with penicillin G. The carrier is soluble in an organic phase, and the Pen G is soluble in an aqueous phase. The Pen G-carrier complex however, is soluble in organic phase. The complex formation reaction occurs at the organic-aqueous interface, and the complex is rapidly transported into the organic phase [3], achieving extraction of the penicillin from the aqueous phase.

Amines are generally used as carriers for reactive extraction. Primary amines have a fairly high solubility in water and causes stable emulsions, so they are not very good as carriers [8]. When secondary and tertiary amines are used as carriers, the extraction is possible within the pH range of 5-7, in which penicillin is stable at room temperature. Secondary amines give the highest degrees of extraction, but can react with the solvent, which is a disadvantage. The distribution coefficients of the complexes of tertiary amines are significantly lower than those of the secondary amines. Quaternary amines are not suitable as re-extraction is often difficult [8].

A secondary amine that is commonly used is Amberlite LA-2 (N-lauryl-N-trialkyl methyl amine) has been chosen as carrier along with other carriers [4, 6, 8, 9, 13].

Penicillin G (HP) is a weak acid (pKa = 2.75). Therefore, the following equilibrium exists when penicillin is dissolved in water

$$HP \longleftrightarrow H^{+} + P^{-} \qquad (1)$$

where H+ and P- represent the hydrogen ion and penicillin acid ion in the aqueous phase, respectively. According to many studies [4, 6, 14, 15], the carrier and penicillin react in a 1:1 ratio, according to the following equation;

$$H^{+}_{(aq)} + P^{-}_{(aq)} + LA_{(org)} \longleftrightarrow HP\text{-}LA_{(org)} \qquad (2)$$

where LA represents the Amberlite-LA2 dissolved in the organic phase (and soluble only in it). However, some more recently published papers [4, 10, 17] have suggested that they react in 2:2 ratio according to the following equation:

$$2HP_{(aq)} + 2LA_{(org)} \longleftrightarrow (LA)_2(HP)_{2\,(org)} \qquad (3)$$

For the reaction mechanism stated above (2:2 ratio), the equilibrium constant (K_{eq}) can be expressed as

$$K_{eq} = C_C / (C_A^2 C_{HP}^2) \qquad (4)$$

where C_C and C_A are the concentrations of amine-complex and amine in the organic phase, respectively and C_{HP} is the concentration of penicillin G in the aqueous phase.

Taking logs of both sides and rearranging, this can be expressed as

$$\text{Log}\,[C_c/C_A^2] = 2\log\,[C_{HP}] + \log \text{Keq} \qquad (5)$$

This mechanism can be verified by plotting log $[C_c/C_A^2]$ versus log $[C_{HP}]$ and determining the slope of the plot.

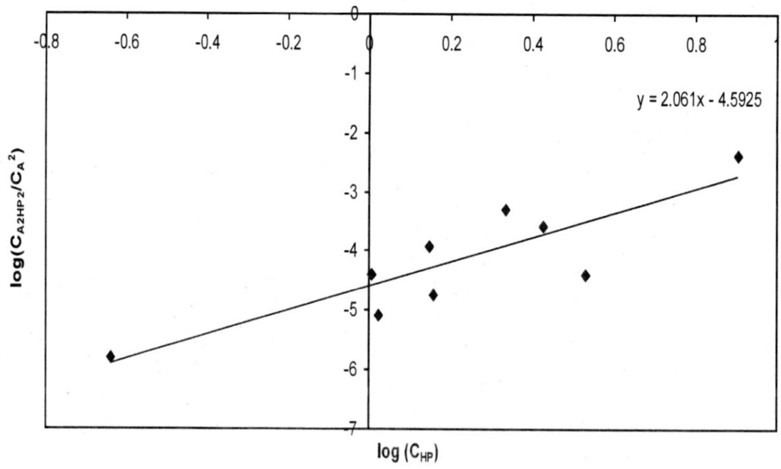

Figure 1. A plot of log $[C_c/C_A^2]$ versus log $[C_{HP}]$.

The experimental data of penicillin G extraction with Amberlite-LA2 are plotted in Figure 1 and a slope of 2 confirms that the reaction takes place with two molecules of penicillin G reacting with two molecules of amine carrier to form a penicillinG-carrier complex. The other data in the literature also suggest this mechanism [27].

The kinetics of penicillin G extraction can be determined from experiments with Amberlite LA-2 dissolved in a single diluent such as kerosene [28] and in a mixture of diluents, e.g. a mixture of kerosene and n-butyl acetate [29]. The experiments reported in the literature, were conducted in a constant-interface area cell and various rate equations were proposed. For the overall reactive extraction the stoichiometry given by equation (2) was used. On the basis of the assumptions of linear concentration gradient in both the aqueous and organic films and applying two-film theory the flux equations were derived. The values of the forward extraction rate constant, $k_1 = 1.64$ l mol^{-2} s^{-1} and the backward rate constant $k_2 = 6.56 \times 10^{-5}$ l s^{-1} were obtained when kerosene was used as solvent. Also it was found that the interfacial-reaction is the rate-controlling mechanism. A mechanism similar to the above was also determined [28] although the extraction system was different and used a mixture of two diluents. These authors determined the intrinsic rate constant and showed that the liquid-liquid reactive extraction of penicillin G is governed by the combined diffusion of the complex (of penicillin G and amine) in the organic layer and penicillin G in the aqueous layer. They also studied the re-extraction from the organic phase using aqueous solution of sodium carbonate as strippant and showed that this is controlled by dissociation reaction of the complex.

Extraction equilibria of penicillin G in various types of organic solvent systems can be examined as reported in [27]. The solvents were: (1) amphiprotic type such as 1-octanol and 1-hexanol, (2) hydrogen-bond acceptor type suc as n-butyl acetate and di-n-butyl ether, (3) hydrogen-bond donor type such as chloroform and (4) apolar aprotic inert solvents such as n-heptane and cyclohexane. It was observed that the solute-solvent interaction was the most effective in physical extraction with the first three types of solvent. From the experimental results for extraction with Amberlite-LA2, it was verified that the penicillin G-amine complex are in the stoichiometric ratio of 2:2 and 1:1 for nonpolar solvents and hydrogen-acceptor solvents, respectively. For other two solvents the stoichiometry of the reactive extraction of penicillin G and amine can be expressed by non-natural numbers.

Physical and Reactive Extractions of Penicillin G

Equilibrium Experiments

For all equilibrium experiments, 5ml organic solution (carrier dissolved in the desired solvent) is put into a centrifuge tube with a pipette, 5ml of prepared penicillin in buffer solution is then added to the tube. This is vortexed for two minutes, then centrifuged after phase separation. This procedure gives accurate results at higher pH because at pH lower than 2 this might cause decomposition of penicillin G if the preparation time is too long. The tubes are then centrifuged at 4000 rpm for 15 minutes in order to completely separate the two phases. The bottom aqueous layer is removed with an eppendorf pipette, and diluted with an appropriate amount of buffer, depending on the initial concentration, in order to measure the absorbance accurately. The concentrations of Penicillin G in the samples are measured using a UV-VISIBLE spectrophotometer (Perkin Elmer 35 UV/VIS spectrophotometer, USA) at a wavelength of 257 nm. More accurate measurements HPLC method is recommended.

The effect of fermentation media is studied by preparing pure Penicillin G either in the media. A typical composition of the fermentation media and the biomass characteristics (as supplied by Rort Richard Laboratories, Auckland, New Zealand) are given in Table 1.

The equilibrium between the aqueous phase and the organic phase is expressed by an apparent distribution coefficient (D_E). The value of this coefficient is expected to depend on the feed solution concentration and pH, carrier type, diluent type and, composition of the organic phase.

$$D_E = \frac{C_{Pen\,Tfo}}{C_{Pen\,Tfi}} \tag{6}$$

The results for the physical extraction of penicillin G (in aqueous solution with no fermentation media) in a variety of solvents at feed pH of 4.5, 5.0 and 5.5 are presented in Figure 2. The results show that tributyl phosphate is the best solvent and butyl acetate is the second best. Extraction at pH 4.5 is also higher than those at the higher pHs. This will be discussed further in the following sections. As has been shown previously [30], the distribution coefficient of penicillin G between the solvent and the aqueous phase depends on the polarity of the solvent, and approximately a linear relationship exists between the dipole moment of the solvent and the log of the partition coefficient. Tributyl phosphate, being the most polar solvent that was used, accordingly gives the highest physical extraction, while butyl acetate, the next most polar solvent, gives second best physical extraction. Shellsol TK, olive oil, and oleyl alcohol are all relatively non polar so they gave much lower distribution coefficients.

The effect of the addition of carrier (reactive extraction) gives much higher distribution coefficient compared to those obtained in physical extraction (Figure 3). In general, the values of the distribution coefficient

Table 1. Composition and properties of the fermentation media and biomass

Composition/properties	Concentration
Neopeptone	10 g/L
Dextrose	20 g/L
pH	5.6
Temperature	25 ^{0}C
Density of broth suspension	1.005 g/mL
Solid mycel content	11-15 v/v%

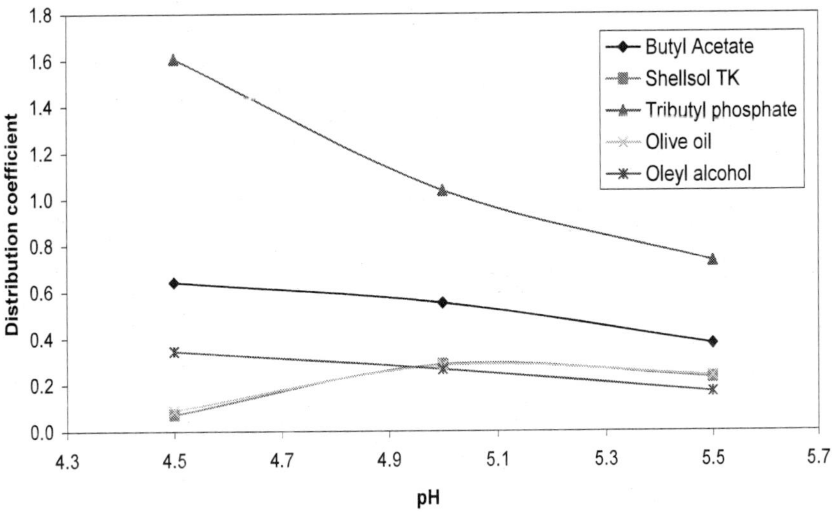

Figure 2. Effect of solvent type at various feed solution pH.

decreases with the increase in feed pH, this effect diminishes at pH higher than 4.8 (shown in Figure 4). This is much pronounced in reactive extraction (Figure 5) compared to the physical extraction (Figure 5) where the effect of pH change is minimal. At higher pH > 4.8 there seems to be no benefit in reactive extraction as the values of the distribution ratio are lower to those obtained by the physical extraction alone. This is because at higher pH the concentration of dissociated penicillin G molecules (that is available for reactive extraction to occur) are less than those at lower pH. However, the distribution coefficient in reactive extraction can be enhanced by increasing concentration of the carrier (shown in Figure 5). While all the distribution coefficients are similar the carrier dioctylamine give slightly higher distribution coefficient in pH range 4.5 – 5. However, the recovery of penicillin G from this phase is rather difficult as reported earlier [8].

The correlation of DE with the feed pH is shown in Figure 4. A decrease in pH means an increase in free H^+ ions in solution. This increases the rate of reverse reaction of equation 1, so more penicillin G is in the form of designated by HP. As shown in equation 3, penicillin G needs to be in this undissociated state for the complexation reaction to occur, so an increase in penicillin in the form of HP increases the rate of the forward reaction, causing formation of more penicillin G - amine complex, which is then extracted in greater proportion to the organic phase. A linear relationship is obtained between the log of the distribution coefficient and the pH, as determined by the following equation;

$$\log (D_E) = -0.34*pH + 2.48 \tag{7}$$

Since all of the solvents studied above are considered to be toxic and do not have any renewable source of supply, vegetable oils (with sunflower oil as a model) were examined as a possible diluent for the extraction process. The experimental results are discussed in the next few pages.

Shown in Figure 6 is a comparison of the various solvents reactive distribution coefficients at a pH of 6. The results also show that tributyl phosphate is able to extract better than both sunflower oil and pegasol (a solvent similar to kerosene). The observations are similar to those noted in previous results that the distribution coefficient for tributyl phosphate increases greatly when Amberlite-LA2 is added to it to allow reactive extraction. Although tributyl phosphate performed best because of its undesirable characteristics such as toxicity, corrosiveness and cost it can not be recommended for industrial scale processing and therefore the experiments were conducted with other solvents.

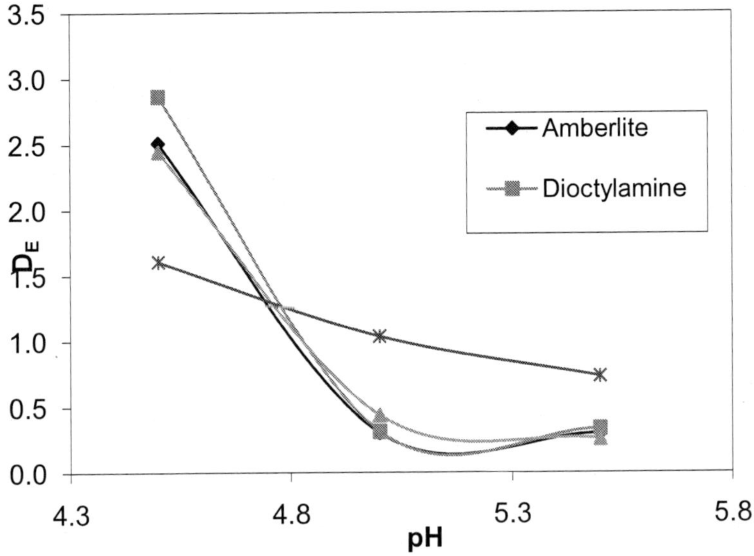

Figure 3. Effect of carrier type at various feed solution pH.

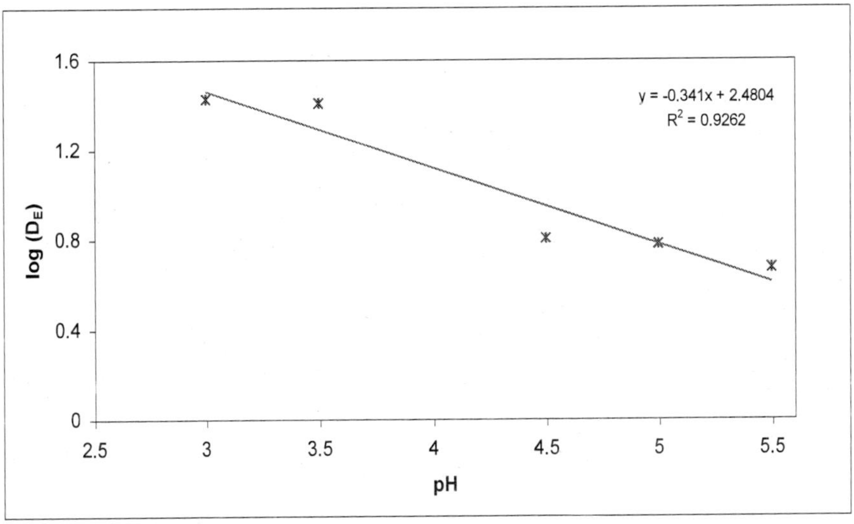

Figure 4. Correlation of log(D_E) versus feed solution pH.

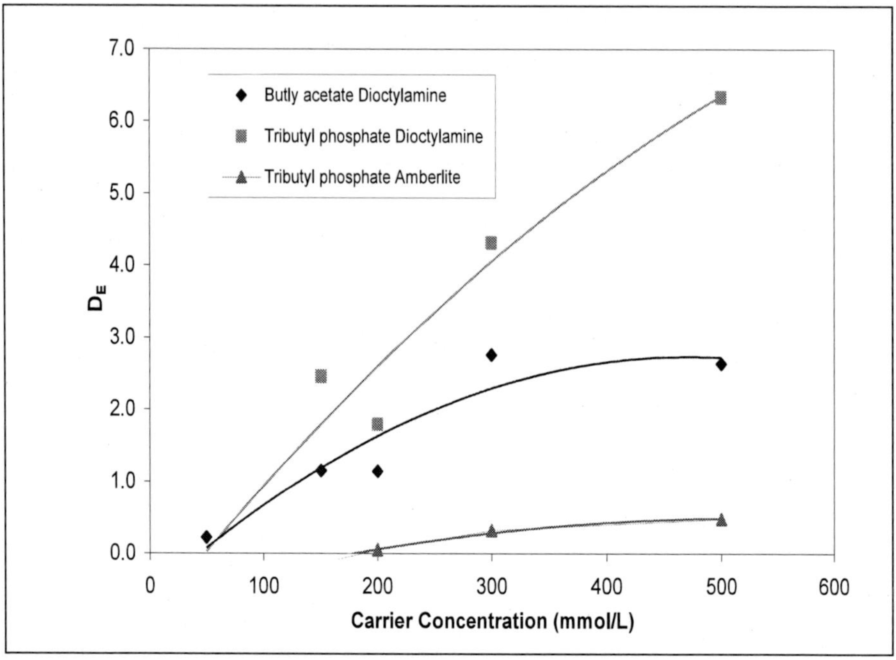

Figure 5. Effect of carrier concentration on reactive extraction.

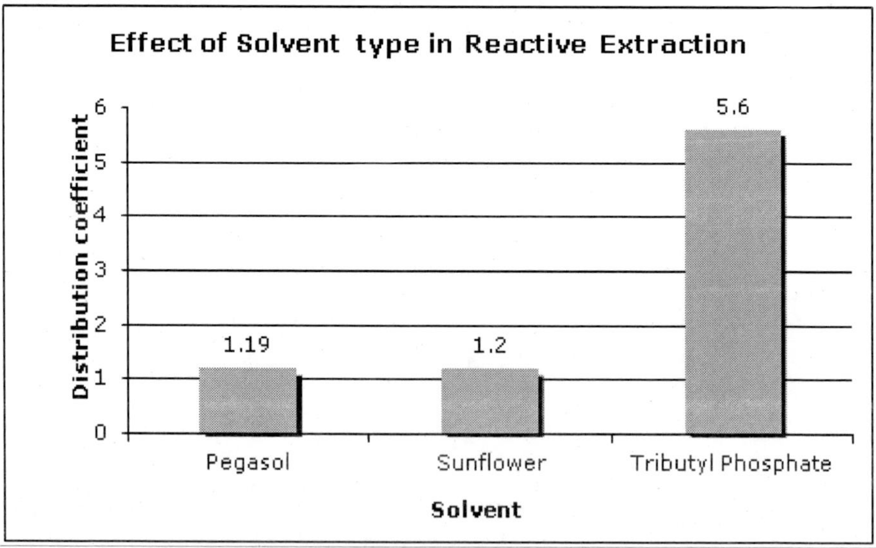

Figure 6. Comparison of Solvents for Reactive Extraction at pH = 7.0 and 5% Amberlite LA-2.

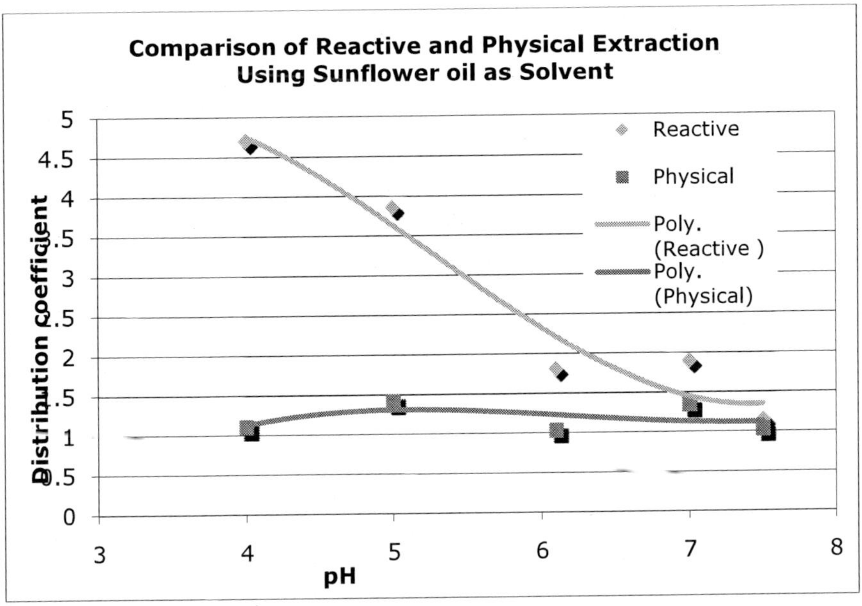

Figure 7. Comparison of reactive and physical extraction using sunflower oil.

The effect of pH on the distribution of penicillin G is shown in Figure 7 for reactive extraction with Amberlite –LA2 in sunflower oil. It is observed that the extraction is highly dependent on pH, with the highest distribution coefficient of 4.8 obtained at a pH of 4. However, at pH values close to fermentations (i.e. 6.5 to 7.5) the difference between physical and reactive extractions are not that significant. The values for reactive extraction are compared to those of Pegasol-Amberlite-LA2 in Figure 8. A similar kind of pH dependence is observed, with a lower value for the highest distribution coefficient of 2.5 at a pH of 4. The value of the distribution coefficient sharply decreases to a low value of 1.3-1.7 at pH 6.5 - 7.5, at which most fermentation is carried out.

Figure 9 shows the effect of Amberlite-LA2 on the extraction at pH 6.5 using sunflower oil.

The results in this graph show that up until a point the distribution coefficient increases with the increase in carrier concentration. The highest distribution coefficient occurred with a carrier amount equal to 5% (v/v) of the total volume of the organic solvent. The trend line suggests that a carrier concentration between 5 and 10% (v/v) would offer the greatest advantage in the extraction of penicillin G.

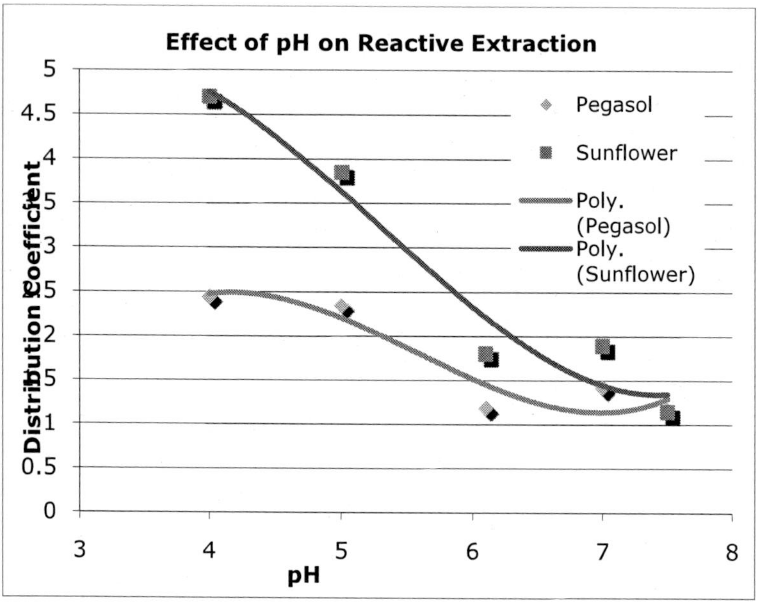

Figure 8. Effect of feed solution pH on reactive extraction with sunflower oil.

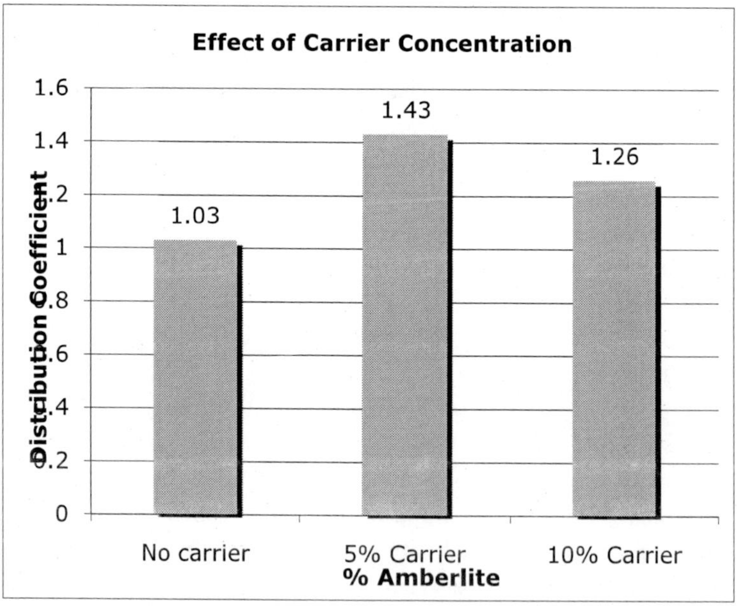

Figure 9. Effect of Carrier concentration on extraction using Sunflower oli-AmberliteLA-2.

The effects of the biomass and fermentation medium have been examined to mimic the effect of a fermentation broth. With the presence of biomedium the value of the distribution coefficient decreases from 1.8 to 1.7, the decrease is more pronounced when the biomass is added (the decrease is from 1.7 to 1.4). The effect of the biomass could be approximated as a diffusion resistance, inhibiting the movement of the carrier and Pen G in the solutions. There may be unseen interactions of the biomass components with the Pen G and further investigations are required to establish and quantify such an effect.

Once the organic phase is selected the process can be upgraded in a pilot-scale module such as hollow-fibre membrane contactor and the extraction performance can be evaluated. In the following section a module is described and its performance is discussed in the extraction of penicillin G.

HOLLOW FIBER MEMBRANE MODULE

A microporous hollow fiber membrane contactor, 5PCM-218, was purchased from Hoechst Celanese Corporation, Charlotte, NC, USA. The contactor has a total of 10,000 polypropylene hollow fibers (Celgard X-30, 240 μm ID, 300 μm, OD) potted with polyethylene epoxy in a polypropylene case of 2.5 cm ID. The characteristics of the contactor are listed in Table 2.

Table 2. Characteristics of the Module

Characteristics of the module	
Fibre Type	Microporous Polypropylene Hollow Fiber
Number of Fibers	10,000
Effective Pore Size	0.05 μm
Effective Fiber Length	15 cm
Porosity	40%
Operating Chracteristics	
Maximum Operating Temperature	60°C
Maximum Operating Pressure	4.1 bar (60 psi)
Cartridge Length	16 cm
Shell Inner Diameter	6.3 cm
Effective Mass Transfer Area	1.4 m^2
Effective Area/Volume	29.3 cm^2/cm^3
Priming Volumes:	
Tubeside (include endcaps)	145 cm^3
Shellside	195 cm^3

Table 3. Feed Solution

Feed Solution	
Concentration of penicillin (mM):	10
Volume (L)	0.4-0.5
pH (-):	4.5
Flow rate (L/h):	14 - 17
Organic solution	
Concentration of carrier in the solvent (mmol/L):	253
Volume (L):	0.5
Flow rate (L/h):	13 - 16

Hollow-Fiber Experiments: Effect of Solvent Type

The experiments were conducted in the recycle mode with both the feed and the organic phases were recirculated. The conditions of the experiments are listed in Table 3. The experiments are conducted with feed and the organic solutions being circulated so that the system behaves as batch systems. The flowrate of the aqueous solution is kept at high flow range, e.g. in the range 13-16 L/h. The operating conditions are kept constant except for the one whose effect is being examined. The results of the effect of three organic solvents are shown in next section with Amberlite LA-2, used as the carrier and shellsol TK, tributyl phosphate and butyl acetate as the solvents, respectively.

The degree of extraction, E(%), for the HFM experiments is measured by the extraction efficiency which is defined by the following equations:

$$E\ (\%) = (C_{penGi} - C_{penGf)/}C_{penGi} \text{ x } 100 \quad (8)$$

where C_{penGi} and C_{penGf} are the initial and final concentration of penicillin G, respectively.

As can be seen from Figure 10, butyl acetate only reached approximately 63% extraction, and this stayed rather steady throughout the remainder of the experiment. Both the organic phases with tributyl phosphate and shellsol TK (a diluent similar to kerosene) however, reached approx. 98% extraction. For tributyl phosphate, it took only 20 minutes to reach the steady state maximum, while for shellsol TK, it took about an hour. The faster initial rate of extraction in tributyl phosphate could be due to its greater distribution coefficient value compared to the organic phase of shellsol TK. After the initial period (time taken to partition into the organic phase) the extraction rate with shellsol TK becomes rapid and eventually attains the same value as that of tributyl phosphate at a longer processing time. This increased extraction rate could be due to greater distribution coefficient at lower pH, lower viscosity (shellsol TK has 10 times smaller viscosity than tributyl phosphate) that reduces the mass transfer resistance once it has partitioned in that phase.

EXTRACTIONS WITH SUNFLOWER OIL

Because of the disadvantages of high toxicity, the cost of processing, the environmental impact and non-renewable sources of supply the other solvents are considered for potential applications. One non-traditional solvent, sunflower oil, is selected as one possible substitute which may overcome some of the disadvantages of the chemical solvents examined in previous studies. The results of extractions with sunflower oil in hollow-fibre modules are discussed in next few paragraphs.

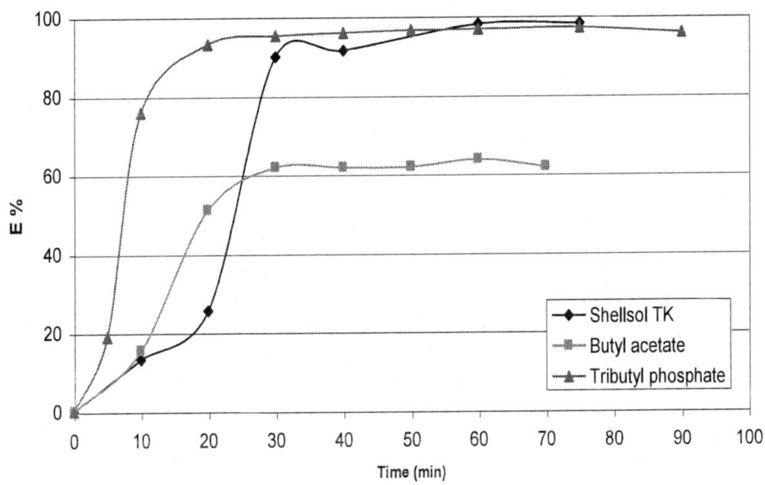

Figure 10. Extraction (%) of penicillin G with various diluent-Amberlite-LA2 as organic phase.

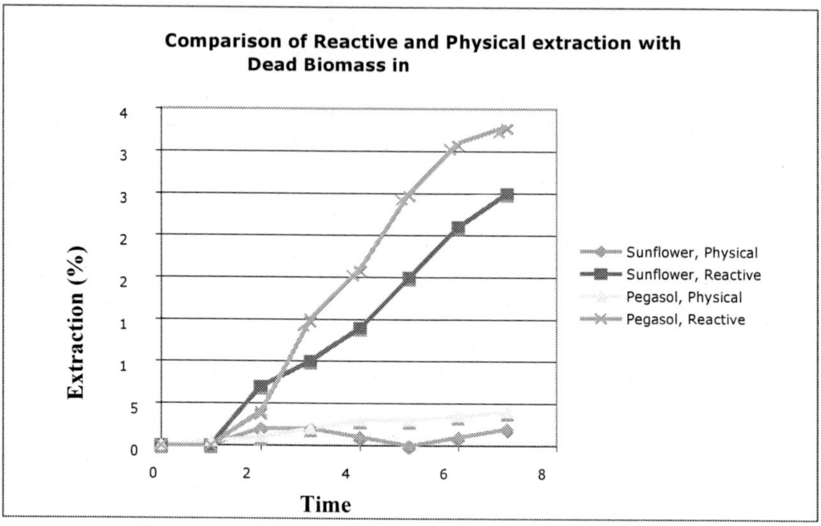

Figure 11. Effect of biomass and biomedium: comparison between physical and reactive extraction.

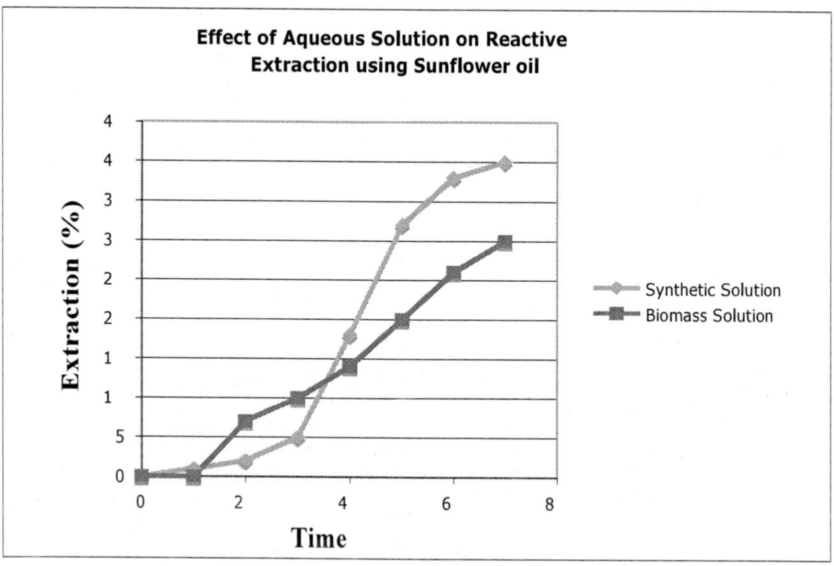

Figure 12. Comparison between synthetic and biomass solution on the extraction of Pen G.

Figure 11 shows the percentage extraction of penicillin G from an aqueous solution (no biomedium and biomass) at pH 6.5 using sunflower oil (the results of pegasol are included for comparison) as the organic phase. Considering the fact that the extractions are performed in a single stage, the percentages obtained are good, approx. 30% with Amberlite LA-2 in sunflower oil and approx. 37% with Amberlite LA-2 with Pegasol within 1 hour of operation. The trend of the plot suggests that further extraction would be achieved by running the experiment over a longer period of time. The physical extraction is shown to be poor, with extracting approx. 4 - 5%. These results are in consistent with the of the equilibrium distribution coefficients presented earlier.

Shown in Figure 12 is the effect of biomass and biomedium compared to that with an aqueous solution (without these components). The percentage extractions are lower due to the presence of the biomass and biomedium. During an extraction period of 70 minutes the percentage extraction dropped from 40% to 30% when sunflower oil was used as an organic phase. This reduction is due to the additional components that could have provided extra resistance to the mass transfer from the bulk solution to the interface. The extent of extraction would have increased if the processing was continued for a longer time. The comparison between the extraction from a solution with biomass and biomedium at pH 6.5 using sunflower oil (5% v/v Amberlite LA-2) is shown in Figure 4. The extraction is still possible and the inhibitory effects of the product could be reduced by some proportion. Thus the productivity can be enhanced when the reactive extraction is operated *in situ* and inhibitory products are removed simultaneously.

Figure 13 below shows the effect of varying penicillin G concentration in the biomass solution on the extraction using sunflower oil (5% v/v Amberlite-LA2). The extraction increased with the increase in Pen G concentration and reached the maximum within the concentration range 20-40 mmol/L. This is expected as the concentration of Amberlite LA-2 was kept constant and the system would have reached saturation at this combination of penicillin G and Amberlite-LA2. For the given conditions there was an optimal concentration of penicillin G (20 - 40 mmol/L) for the highest extraction. Below this concentration, the extraction rate decreased, indicating that carrier was in excess to the relative amount of penicillin. Above the optimum concentration the rate of reactive extraction decreased, indicating the amount of penicillin G was present in excess relative to the amount of carrier, and therefore was acting as a barrier to further extraction.

Re-extraction (Recovery) of Penicillin G from the Organic Phase

For recovery of penicillin G from the oil phase, aqueous solutions of sodium chloride and sodium carbonate were used. The values of the operating conditions are listed in Table 4. Shown in Figure 14 are the values of the re-extraction percentage of Penicillin G from sunflower oil. The results suggest that sodium carbonate is readily able to re extract penicillin from the oil phase, recovering 98% of the penicillin within a reasonable time and using a small concentration. Sodium Chloride was less successful with an extraction of 40% in the same time frame. These results appear to correlate well to previous studies with other solvents [31].

In the next section, mathematical modelling is presented for liquid-liquid extraction and mass transfer processes in a commercially available membrane contactor (i.e. HFMC). Rather than a comprehensive approach a simple analysis is proposed to provide an understanding of the mass transfer phenomena in a small pilot-scale module.

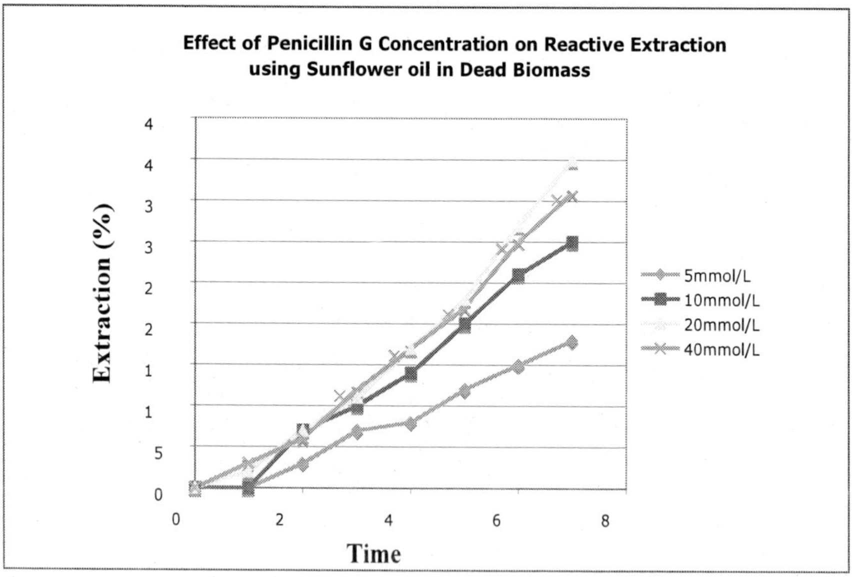

Figure 13. Effect of Pen G concentration in the biomass solution on extraction (%).

Table 4. Recovery and Organic Solution

Recovery Solution	
Concentration of Strippant (mM):	10
Volume (L)	0.4-0.5 (recirculating volume)
pH (-):	4.5
Flow rate (L/h):	13.6 - 14.4)
Organic solution	
Concentration of carrier in the solvent (mmol/L):	253
Volume (L):	0.5 (recirculating volume)
Flow rate (L/h):	13.3 - 14.1

Mathematical Modelling of Mass Transfer in Hollow-fibre Module

The reaction of undissociated penicillin G (HP) with a carrier (LA) dissolved in the solvent gives a reaction complex (HP-LA) which remains largely in the organic phase and may be represented by [9,18, 19, 23]:

A simple 2:2 stoichiometry (the molar ratio of organic acid to that of extractant) has been proposed. However, this depends on the type of the organic acid and its ionic state, the type of the extractant and the type of the solvent [34]. The reaction mechanism does not change the mass transfer processes.

For reactive extractions microporous hollow fiber membrane contactors, in various configurations have been evaluated [14, 33, 34, 35 - 37]. Typically two modules are used, one for the extraction and the other for the re-extraction or stripping process. These are commercially available modules, e.g. a module with catalogue No. 5PCM-218, obtained from Separation Products Division, Hoechst Celanese Corporation, Charlotte, NC, USA, have been extensively used for mass transfer operation. The contactor has a shell-and-tube configuration with a total of 10,000 polypropylene hollow fibers (Celgard X-30, 240 μm ID, 300 μm, OD, length 15 cm) potted in polyethylene in a polypropylene case of 6 cm ID. The surface area of the contactor was 1.4 m^2. The hollow fiber module is usually set up as shown in Figure 15. When these modules are used, aqueous and organic solutions flow continuously, one through the lumen side of the fibre and the other through the shell side. The both the solutions get into contact through the pores of the wall. Phase entrainment is avoided by applying a little higher pressure on the aqueous side. The pressure difference between the phases is between 0.2 – 0.3 bar and it has been reported that it has no influence on the mass transfer processes.

These contactors have been used in various process configurations such as hollow fibre contained liquid membranes, HFCLM [38], hollow fibre supported liquid membrane, HFSLM, [39], non-dispersive solvent extraction, NDSX, [40] and hollow fibre renewal liquid membranes, HFRLM, [41]. The main difference between these configurations is that the contacting pattern of the liquid phases is different. In a newly developed mass transfer operation known as emulsion pertraction, PERT, a single module is used for extraction and stripping simultaneously [33, 37]. The emulsion consists of an organic solvent with a dissolved extractant phase with droplets of strip liquid dispersed in it. The phases are separated by the hydrophobic membrane surface. The contact between the water phase and emulsion phase occurs at the pore mouth. The solute-extractant complex diffuses through the pores and on the other side of the membrane the extractant is regenerated by strip liquid. The analysis below is not applicable to the PERT process it is devoted to extraction in a single module.

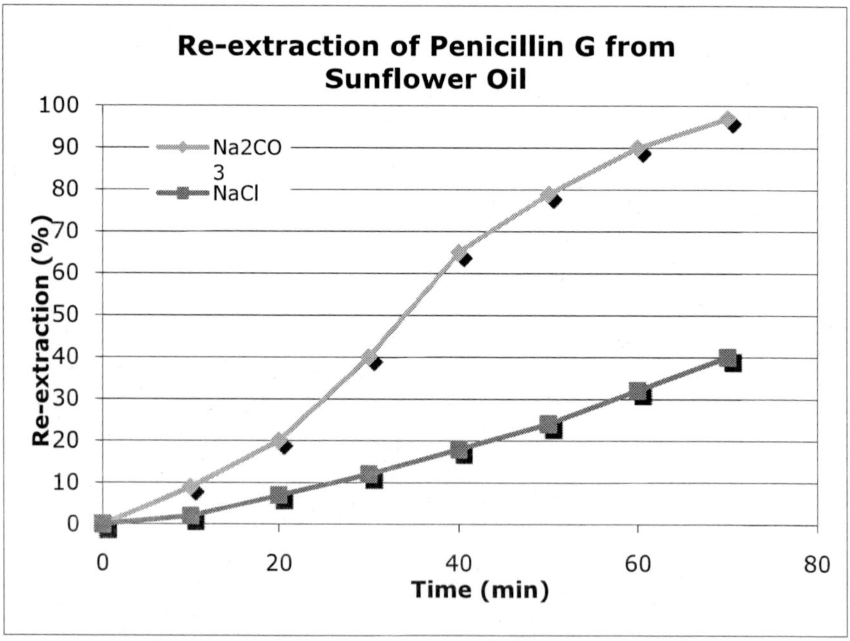

Figure 14. Recovery of penicillin G from the organic phase of sunflower oil.

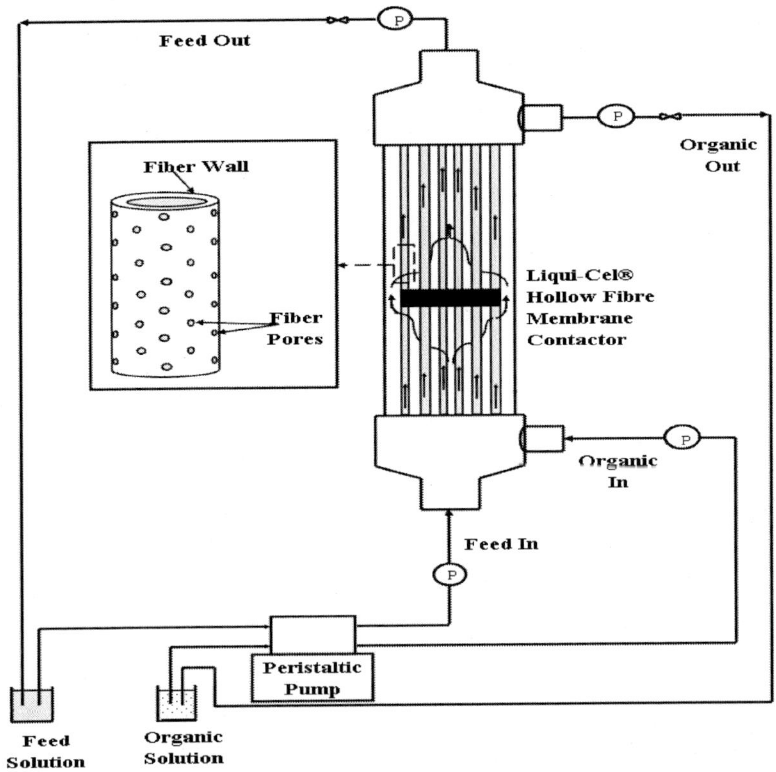

Figure 15. A schematic diagram of the mass transfer operation of the hollow-fibre membrane contactor.

The transport mechanism of solute molecules (penicillin G) from an aqueous feed side to the organic side through hollow-fibre wall can be described by the steps as shown in Figure 16. The mass transfer processes proceed by the solute transport through the resistances from the aqueous feed (inside the fibre) to the organic phase (shell side). The steps considered for the mass transport and reactive extraction are:

- the solute molecules are transported from the feed solution to the feed-pore interface and can be expressed by Eq. (9).
- at the interface the reaction between the solute and the carrier takes place (Eq.1) to form a solute-carrier complex. The equilibrium

concentrations at the interface in the aqueous and organic phases can be related by an apparent distribution coefficient (DE), given by Eq. (6).

- the apparent diffusion of the solute-carrier complex through the pores of the hollow-fibers filled with the organic phase and this can be expressed by Eq. (10).
- the final step is the transport through the solvent boundary layer at the outer end of the pore mouth and this step can be expressed as in Eq. (11).

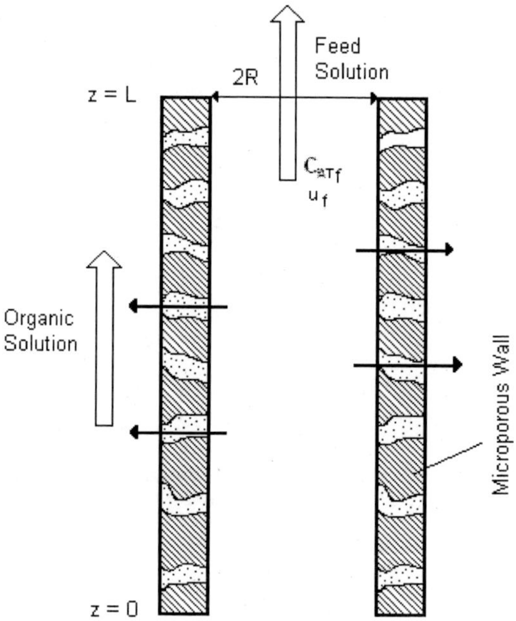

$$\text{Convective flux} = -u_f\left(\frac{V}{A}\right)_{in}\frac{\partial C^m_{HPf}}{\partial z}$$

The interfacial reaction :

$$(HP)_{aq} + (LA)_{org}^{-} \Leftrightarrow (HP - LA\ )_{org}$$

$\left(HP - LA^{-}\right)$ diffuses through the porous wall

to the shell side (the diffusive flux)

$$\text{Flux} = -K_{of}\left(C^m_{HPf} - \frac{C^m_{HPo}}{D_E}\right)$$

Figure 16. (a) A schematic of the mass transfer processes in the membrane module.

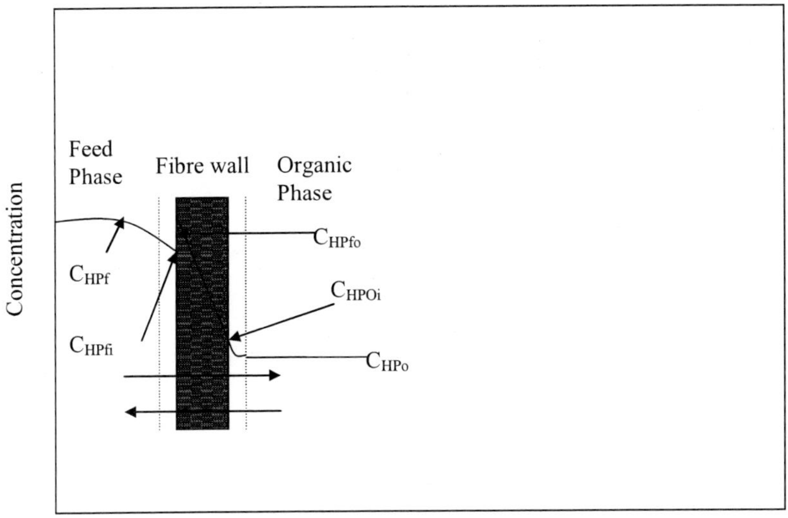

Figure 16. (b) Concentration profiles of penicillin G in the feed phase (inside the fibre), in the membrane wall and in the organic phase (inside the shell).

The following assumptions have been considered for writing the model equations:

- The system works at isothermal conditions.
- Equilibrium is reached at the interfaces of the aqueous and organic phases.
- The curvature of the interfaces does not affect significantly the processes.
- The distribution coefficient of the solute is considered to be constant with the conditions used
- Uniform pore size along the entire length of the contactor.
- The mass transfer processes in the boundary layer is described by the diffusion model.
- The phases are immiscible and the pores are wetted by the organic phases only.

The following equations describe the mass transfer and reactive processes

$$N_{HPf} = k_f \left(C_{HPf} - C_{HPfi} \right) \tag{9}$$

$$N_{HPM} = k_{mf}\left(C_{HPfo} - C_{HPOi}\right) \tag{10}$$

$$N_{HPo} = k_o(C_{HPOi} - C_{HPO}) \tag{11}$$

where k_f, k_{mf} and k_o are the mass transfer coefficient in the feed side, on the membrane and in the organic side, respectively. The concentrations C_{HPf} and C_{HPfi} are the total penicillin G concentrations in the bulk aqueous solution and at the feed-membrane interface, respectively. The concentrations C_{HPOi} and C_{HPO} are the concentration of penicillin G at the membrane-organic interface and in bulk organic phase, respectively. In the above equations, the concentration denoted by $(C_{HP(org)})$ has been taken as the total penicillin G concentration in the organic phase and denoted by C_{HPo} .

Combining the above equations the flux of SJM at steady state is obtained as:

$$N_{HP} = K_{of}\left(C_{HP(aq)} - C_{HP(org)}/D_E\right) \tag{12}$$

where K_{of} is the overall mass transfer coefficient of the process. The concentrations $C_{HP(aq)}$ and $C_{HP(org)}$ are the total Pen G concentrations in the bulk solution and at the feed-membrane interface, respectively.

K_{of} is also related to the individual transfer coefficients and various correlations are available in the literature. In order to calculate K_{of} from the above equation the individual mass transfer coefficients have to be known in addition to the distribution ratio of Pen G. There are many correlations available in the literature for calculating the individual mass transfer coefficients. Each is based on the specific experimental conditions and equipment set-up used to develop the correlation. So the validity of the correlation needs to be matched for its appropriateness before applying to any other system.

In the section below the mathematical model and its solution for recycling mode operation is presented here and the details are available in an earlier paper (16). This consists of the following equations, Eq. (13) and (14) that defines (i) the change in Pen G concentration in the module and (ii) the change in Pen G concentration in the feed reservoir, where aqueous solution is continuously circulated.

Mass balance in the hollow-fiber module

$$\left(\frac{V}{A}\right)_{in} \frac{\partial C^{m}_{HP(aq)}}{\partial t} = -u_f \left(\frac{V}{A}\right)_{in} \frac{\partial C^{m}_{HP(aq)}}{\partial Z} - K_{of}\left(C^{m}_{HP(aq)} - \frac{C^{m}_{HP(org)}}{D_E} \right) \quad (13)$$

Mass balance in the reservoir

$$\frac{dC^{t}_{HP(aq)}}{dt} = \frac{q_f}{v^t}\left(C^{m}_{HP(aq)}\Big|_{z=L} - C^{m}_{HP(aq)}\Big|_{z=0} \right) \quad (14)$$

where $\left(\frac{V}{A}\right)_{in}$ is the ratio of the volume to inner area of mass transfer in the fibers, L is the length of the fiber, u_f is the linear velocity, q_f is the feed flow rate and v^t is the reservoir volume. The superscripts m and t refer to the membrane module and tank, respectively.

The ratio, $\left(\frac{V}{A}\right)_{in}$, for the feed solution circulating along the inside of the fiber is

$$\left(\frac{V}{A}\right)_{in} = \frac{(\pi di^2 L/4)\, nf}{(\pi\, diL)\, nf} = \frac{di}{(4)} \quad (15)$$

where di is the inner radius and n_f is the number of hollow fibers in the module.

The factor, $\left(\frac{V}{A}\right)_{in} \frac{1}{L}$, is a small number for this type of contactor, i.e. 4 x 10^{-4}. Using this and assuming a slow rate of change of solute concentration in the module, the model can be solved analytically and the final form of the solution is:

$$\mathrm{Ln}\left\{\frac{C_{HPi(aq)}}{(1+1/D_E)\,C_{HPf(aq)}-\dfrac{C_{HPi(aq)}}{D_E}}\right\}=\frac{q_f}{v^t}\left\{1-\exp\left(-BK_{of}\right)\right\}t \tag{16}$$

where B has been defined by the following equation:

$$B=\frac{\left(1+\dfrac{1}{D_E}\right)L}{\left(\dfrac{V}{A}\right)_{in}u_f} \tag{17}$$

The overall mass transfer coefficient (K_{of}) can be determined from the value of the slope of the linear plots of the left hand side (LHS) of equation (16) versus t (time). From the slope of the line and calculating B from equation (17) (using the velocity in side the fibre, module volume, mass transfer area and the distribution coefficient), the value of K_{of} can be estimated.

Various sets of extraction data of penicillin G can be plotted as LHS of Eqn. (16) versus process time. The values of the overall transfer coefficient (K_{of}) are estimated from the slope of such plot. The value of the distribution coefficient for the respective organic phase is required in the calculation. The experimental data during the initial period (approx. 30-40 minutes) shows good correlation (R^2 = 0.8) and are used in the plot. From the best fit line of the plot of LHS versus time (Eq. 16), the slope is obtained and the value of the overall mass transfer coefficient (K_{of}) is estimated. The calculated values of the (K_{of}) are in the range 0.4 - 5.6 x 10^{-5} cm/s and they are listed in Table 5 for various organic phases (all containing Amberlite - LA2 at the same concentration). This is consistent with the distribution coefficient values (Figure 3) and the extraction data (Figure 7), i.e. the mass transfer rate is much faster with tributyl phosphate than the other two organic phases. The values reported in the literature (8) for extraction of penicillin G are in the range: (3.6-11.1) x 10^{-5} cm/s. Considering the fact that their experiments (8) were carried out for simultaneous extraction and re-extraction at the optimized operating conditions, the values obtained in the present report compare well. It is worthwhile to mention the experiments in Ref 8 were performed with kerosene, a hazardous and toxic solvent from the points of environment, operator safety and handing. Also it creates swelling problems in hollow-fiber membrane modules affecting the process performance.

Moreover the mass transfer coefficient has been estimated using a semi-analytical solution and uses the values determined experimentally in this membrane contactor. Furthermore, this estimation does not require any correlations or data or parameter from previous literature. The knowledge of the analysis of results in terms of mass transfer coefficient could be used to estimate of the membrane area for a large-scale production process. In order to gain better understanding, a comprehensive model (with various mass transfer resistances) has to be solved with relevant correlation for the extra-flow hollow-fibre module and for the experimental values of the parameters of the system considered.

As mentioned earlier in Equation (12), the overall mass transfer coefficient can also be calculated using the individual mass transfer coefficients: mass transfer coefficient in the aqueous side (k_f), membrane mass transfer coefficient (k_{mf}) and the mass transfer coefficient in the organic side (k_o). In most of the hollow fibre systems, the aqueous phase containing the "target" component is allowed to flow (under laminar conditions) through the inside of the hollow fibres. The mass transfer coefficient in the aqueous side can be calculated using an equation of the following form [42]:

$$k_i d_i/D_p = 1.62 \text{ x } (d_i^2 v_i/LD_p)^{0.33} \qquad (18)$$

where di is the diameter of the fiber, Dp is the diffusivity of the penicillin G molecules, v_i is the feed velocity in the fibers and L is the fiber length. The values of k_i reported [19] within the Reynolds number range (0.08-200) are 1.8 x 10^{-6} m/s to 9.5 x 10^{-4} m/s.

The membrane mass transfer coefficient (k_{mf}) is calculated from the following equation [34, 43]:

$$k_{mf} = D_X/\sigma_M \, (\varepsilon/\tau) \qquad (19)$$

where D_x is the diffusion coefficient of the Amberlite LA2-penicillin G complex in the organic phase, σ_M is the membrane thickness, ε is the porosity and τ is the totuosity. The value of D_X in the organic phase is estimated by the following correlation [47]:

Table 5. Concentration versus Feed

Amberlite- LA2 concentration in the solvent, and organic phase flow rate, OPR	Feed concentration (Penicillin G) in aqueous solution or fermentation media	Feed flow rate, FFR and the ratio of flow, RF)	Overall mass transfer coefficient (K_{of}) cm/s x 10^5
Butyl acetate (solvent) = 5% v/v and OPR = 200 ml/min	100 mmol/L in aqueous solution	FFR = 230 ml/min RF = 1.15	1.7
Shellsol TK (solvent) = 5% v/v and OPR = 200 ml/min	100 mmol/L aqueous solution	FFR = 230 ml/min RF = 1.15	1.8
Tributyl phosphate (solvent) = 5% v/v and OPR = 200 ml/min	100 mmol/L in aqueous solution	FFR = 230 ml/min RF = 1.15	5.6
Sunflower oil (solvent) = 5% v/v and OPR = 180 ml/min	20 mmol/L in fermentation media	FFR = 200 ml/min Ratio of flow 1.11	0.4
Sunflower oil (solvent) = 5% v/v and OPR = 180 ml/min	20 mmol/L In fermentation media	FFR = 200 ml/min RF = 1.11	0.9
Sunflower oil (solvent) = 5% v/v and OPR = 180 ml/min	20 mmol/L In fermentation media	FFR = 200 ml/min RF = 1.15	1.2

It is noted that the equation will be different (if the organic flow is allowed through the fibres) as reported [48].

$$D_X = 8.2 \times 10^{-8} [1 + (3Vs/Vx)] T/(\mu_S Vx) 0.66 \quad (20)$$

The values of k_{mf} can be estimated [19] and the reported values of D_m are1.5 x 10^{-5} m/s using D_x as 1.1 x 10^{-8} m/s to. Other equations have also been used by researchers in this field [44, 45].

For the phase in perpendicular flow through the shell side of the contactor, the value of (k_o) can be estimated using the following equation [49]:

$$k_o d_o/D_X = 1.4 \times (d_o v_o/D_X)^{0.33} \quad (21)$$

where d_o is the outside fiber diameter and vo is the superficial velocity.

A detailed discussion of various systems and the estimated values of the mass transfer coefficients are available [19]. It is concluded that the estimation by any empirical correlation mentioned in the literature require a description of the specific system, the characteristics of the fibers and contactor, the physical and transport properties of the system under study, the operating conditions and the assumption that the specific system, for which the mass transfer coefficients are being estimated, is at least similar to the one where the correlation was developed.

CONCLUSION

The extraction of penicillin G is accomplished using (i) solvent alone (physical extraction) and (ii) carrier-solvent combinations (reactive extraction). The equilibrium reactions between penicillin G and Amberlite-LA2 (carrier) can be modelled and validated for reactive extractions. The process has been upgraded to a membrane contactor to achieve good percentage extraction. A simplified model solution has also been presented to determine the overall mass transfer co-efficient of the process.

The extent of physical extraction of penicillin G increases with the polarity of the solvent. Active solvent such as tributyl phosphate alone can give better distribution at higher pH range of feed solution (higher than pH

4.9). The reaction between the undissociated penicillin G and carrier molecules follows a 2:2 stoichiometry and it is evident from the experimental results.

Reactive extraction with the carriers like dioctylamine and Amberlite-LA2 in tributyl phosphate/shellsol TK/butyl acetate also gives good values of the distribution co-efficient, especially at pH < 4.8. However, Amberlite-LA2 is preferred as it is less costly and subsequent stripping of penicillin G from this phase is easier compared to the other carrier-solvent combinations.

An extraction of approx. 98% was achieved in an hour using Amberlite-LA2 with these of the solvents: shellsol TK, tributyl phosphate and butyl acetate at a recirculation flow rate of 14 – 17 L/h in a hollow-fibre membrane contactor. Because of the undesirable characteristics such as toxicity, corrosiveness, cost and operational health and safety, these solvents are not recommended for industrial scale processing. The extractions with sunflower oil are not as good as these solvents but have the benefit of overcoming the above-mentioned disadvantages with potentially renewable sources of supply. The conclusions from sunflower extractions are:

- the physical extraction of penicillin G is low compared to the chemical solvents.
- the sunflower oil with many the characteristics of low cost, non-toxic and lower environmental impact, was found to give good values of distribution coefficient and percentage extraction at natural pH of fermentation when carrier was added to it (reactive extraction).
- the percentage extraction of penicillin G at a pH close to fermentation was moderate, approx. 35 - 40% extraction was achieved in a small pilot-scale hollow-fibre membrane contactor and the concentration of Amberlite-LA2 (carrier) required was small (only 5% v/v in sunflower oil).
- Addition of biomass and biomedium to the feed solution was shown to act as a resistance to the process, reducing the extraction to 30 and 35% using only 5% (v/v) Amberlite-LA2 in sunflower oil.
- a good recovery (approx. 96-98%) of penicillin G from sunflower oil phase was possible using an aqueous solution of sodium carbonate.
- the overall mass transfer coefficient can be easily determined from an approximate solution presented in this chapter and the values compare very well with the literature value.

ACKNOWLEDGMENTS

The financial assistance of the United Arab Emirates University, Al Ain, UAE, is gratefully acknowledged. Also thanks to Dean Jones and David Tucker (undergraduate students from the University of Auckland, New Zealand) for performing the experiments.

REFERENCES

[1] B. L. Ligon, B. L. (2004). Penicillin: *Its discovery and early development, Seminars in Pediatrics Infectious Diseases,* 15, 52-57.

[2] R. Sykes, Penicillin: *From Discovery to Product. Bulletin of the World Health Organisation,* 2001.

[3] Matsumoto, M., Ohtani, T., Kondo, K. (2007). Comparison of solvent extraction and supported liquid membrane permeation using an ionic liquid for concentrating penicillin G, *J. Membr. Sci.,* 289, 92-96.

[4] Jiang, Y., Xia, H., Yu, J., Guo, C., Liu, H. (2009). Hydrophobic ionic liquids-assisted polymer recovery during penicillin extraction in aqueous two-phase system, *Chem. Eng. J.,* 147, 22-26.

[5] Liu, Q. F., Yu., J., Liu, H. Z. (2004). Extraction of penicillin G by aqueous two-phase system of ionic liquid [Bmim]BF4 and NaH2PO4, *Chinese Science Bulletin,* 50, 1582-1588.

[6] Liu, Q., Yu., J., Li, W., Hu, X., Xia, H., Liu, H., Yang, P. (2006). Partitioning behavior of penicillin G in a aqueous two-phase system formed by ionic liquids-and phosphate, *Sep. Sci. Technol.,* 41, 2849-2858.

[7] Juang, R.-S., Lee, S.-H., Shiau, R.-C. (1998). Carrier-facilitated liquid membrane extraction of penicillin G from aqueous streams, *J. Membr. Sci.,* 146, 95-104.

[8] Lazarova, Z., Syska, B., Schugerl, K. (2002). Application of large-scale hollow fibre membrane contactors for simultaneous extractive removal and stripping of penicillin G, *J. Membr. Sci.,* 202, 151 – 164.

[9] Reschke, M., Schügerl, K. (1984b). Reactive extraction of penicillin II: distribution coefficients and degrees of extraction, *Chem. Eng. J.,* 28, B11-20.

[10] Lee, S. C. (2004). Kinetics of reactive extraction of penicillin G by Amberlite LA-2 in kerosene, *AIChE J.,* 50, 119-126.

[11] Likidis, Z, Schlichting, E., Bischoff, L., Schügerl, K. (1989). Reactive extraction of Penicillin G from micelle-containing broth in a countercurrent extraction decanter, *Biotechnol. Bioeng.,* 33, 1385 – 1392.

[12] Lee, S. C. (2008). Development of a more efficient emulsion liquid membrane system with a dilute polymer solution for extraction of penicillin G, *J. Ind. Eng. Chem.,* 14, 207-212.

[13] Mohd-Setapar, S. H., Wakeman, R. J., Tarleton, E. S. (2009). Penicillin G solubilization into AOT reversed micelles, *Chem. Eng. Res. Design,* 87, 833-842.

[14] Bi, P. Y., Dong, H. R., Guo, Q. Z. (2009). Separation and purification of penicillin G from fermentation broth by solvent sublation, *Sep. Purif. Technol.,* 65, 228-231.

[15] Bi, P. Y., Li, D.-qing, Dong, H.-r. (2009). A novel technique for the separation and concentration of penicillin G from fermentation broth: aqueous two-phase flotation, *Sep. Purif. Technol.,* 69, 205-209.

[16] Wu,Y., Li, W., Gao, H., Li, Q., Li, Y., Wang, K., Mahmood, I., Liu, H. (2010). A novel continuous re-extraction procedure of penicillin G by a micro-extractor based ceramic membrane, *J. Membr. Sci.,* 347, 17-25.

[17] Wang, S. S.; Lee, C. J. (1995). Kinetics of penicillin G extraction by Amberlite LA-2 as a mobile carrier in a constant-interface-area cell, *Chem Eng. J.,* 58, 285-290.

[18] Lee, S. C., Ahn, B. S. Kim, J.-G. (2002). Reaction equilibrium of penicillin G with Amberlite LA-2 in a nonpolar organic solvent, *Biotechnol. Prog.,* 18, 108-114.

[19] Bringas, E.; San Roman, MF. (2009). An overview of the mathematical modelling of liquid membrane separation processes in hollow fibre contactors. *J. Chem Tech Biotechnol.,* 84, 1583-1614.

[20] Canari R.; Eyal. A. M. (2003). Selectivity in monocarboxylic acids extraction from their mixture solutions using an amine-based extractant: effect of pH. *Ind. Eng. Chem. Res.,* 42, 1307- 1314.

[21] Coelhoso, I.M.; Cardoso, M.M.; Viegas, R.M.C.; Crespo, J.P.S.G. (2000). Transport mechanisms and modelling in liquid membrane contactors. *Sep. Purif. Technol.,* 19, 183-197.

[22] Hossain, Md. M.; Dean, J; (2008). Modelling and analysis of reactive extraction of penicillin G in a hollow-fibre membrane module. *Sep. Purific. Technol.,* 62(2), 437-443.

[23] Datta, R.; Henry, M. (2006). Lactic acid: Recent advances in products, processes and technologies – a review. *J. Chem. Tech. Biotech,* 81, 1119-1129.

[24] Lee, S. C. (2008). Development of a more efficient emulsion liquid membrane system with a dilute polymer solution for extraction of penicillin G, *J. Ind. Eng. Chem.,* 14, 207-212.

[25] Ren, Z.; Zhanh, W.; Li, H.; Lin, W. (2009). Mass transfer characteristics of citric acid extraction by hollow fibre renewal liquid membrane. *Chem. Eng. J.*, 146, 220-226.

[26] Hano, T., Matsumoto, M., Ohtake, T. (1994). "Continuous extraction of penicillin G with liquid surfactant membrane using vibro mixer", *J. Membr. Sci.* 93, 61–68 (1994).

[27] Wang, S. S.; Lee, C. J. (1995). Kinetics of penicillin G extraction by Amberlite LA-2 as a mobile carrier in a constant-interface-area cell. *Chem Eng. J.,* 58, 285-290.

[28] Lee, S. C. (2009). Extraction equilibria of penicillin G in four different types of organic solvent systems, *J. Ind. Eng. Chem.,* 15, 403-409.

[29] Reschke, M. Schügerl, K. (1984a). Reactive extraction of penicillin I: stability of penicillin G in the presence of carriers and relationships for distribution coefficients and degrees of extraction, *Chem. Eng. J.,* 8, B1 – 9.

[30] Lee, S. C., Chang, J. H, Ahn, B. S., Lee, W. K. (1998). Mathematical Modeling of Penicillin G Extraction in an Emulsion Liquid Membrane System containing only a Surfactant in the membrane phase, *J. Membr. Sci.,* 149, 39-49.

[31] Juang, R.-S., Lin, Y.-S. (1996). Distribution equilibrium of penicillin G between water and organic solutions of Amberlite LA-2. *Chem. Eng. J.,* 62, 231-236.

[32] Klaassen, R.; Feron, P. H. M.; Jansen, A. E. (2005). Membrane contactors in industrial applications. *Chem. Eng. Res. Des.,* 83, 234-246.

[33] Prasad, R.; Sirkar, K.K. (1988). Dispersion-free solvent extraction with microporous hollow-fiber modules. *AIChE J.,* 34, 177-188.

[34] Kertesz, R.; Schlosser, S. (2005). Design and simulation of two phase hollow fibre contactors for simultaneous membrane based solvent extraction and stripping of organic acids and bases. *Sep. Purif. Technol.,* 41, 275-287.

[35] Lin, S.-H.; Chen, C.-N. (2006). Simultaneous reactive extraction separation of amino acids from water with D2EHPA in hollow fiber contactors. *J. Membr. Sci.,* 280, 771-780.

[36] Klaassen, R.; Jansen, A. E. (2001). The membrane contactor: environmental applications and possibilities. *Environ. Prog.,* 20, 37-43.
[37] Dai, X. P.; Yang, Z. F.; Luo, R. G.; Sirkar, K. K. (2001).Lipase-facilitated separation of organic acids in a hollow-fibre contained liquid membrane module. *J. Membr. Sci.*, 171, 183-196.
[38] Rathore, N. S.; Sonawane, J. V.; Kumar, A.; Venugopalan, A. K.; Singh, R. K.; Bajpal, D. D. (2001). Hollow fibre supported liquid membrane: a novel technique for separation and recovery of plutonium from aqueous acidic wastes. *J. Membr. Sci.*, 189, 119-128.
[39] Yang, Q.; Kochergnisky, N. M. (2006). Copper recovery and spent ammoniacal etchant regeneration based hollow fibre supported liquid membrane technology: from bench-scale to pilot-scale tests. *J. Membr. Sci.*, 286, 301-309.
[40] Ren, Z.; Zhanh, W.; Li, H.; Lin, W. (2009). Mass transfer characteristics of citric acid extraction by hollow fibre renewal liquid membrane. *Chem. Eng. J.*, 146, 220-226.
[41] Skelland, A.H.P., Diffusional Mass Transfer, Wiley, New York (1974).
[42] Schlosser, S; Sabolova, E. Mass-transfer characteristics of hollow-fibre contactors for pertraction and membrane based extraction of organic acids. *Proceedings of Engineering with Membranes,* Vol I, paper No. PA.4-17, p313, June 3-6, 2001, Granada, Spain.
[43] de Haan, A.B.; Bartels, P.V.; de Graauw, J. (1989). Extraction of metals from waste water: Modeling the mass transfer in a supported liquid membrane process. *J. Membr. Sci.*, 281-297.
[44] Hu, S.B.; Wiencek. (1998). Emulsion-liquid-membrane extraction of copper using a hollow-fiber contactor. *AIChE J.*, 44, 570-581.
[45] Lee, S. C., Ahn, B. S. Kim, J.-G. (2002). Reaction Equilibrium of Penicillin G with Amberlite LA-2 in a Nonpolar Organic Solvent, *Biotechnol. Prog.,* 18, 108-114.
[46] 46. Reid, R. C., Prausnitz, J. M., Sherwood, T. K. (1987). *The properties of gases and liquids,* 4th Edition, McGraw-Hill, New York, 1987.
[47] Lin, S.-H.; Chen, C.-N. (2006). Simultaneous reactive extraction separation of amino acids from water with D2EHPA in hollow fiber contactors. *J. Membr. Sci.*, 280, 771-780.
[48] Yang, C., Cussler, E. L. (2000). Reactive Extraction of Penicillin G in Hollow-Fiber and Hollow-Fiber Fabric Modules. *Biotechnol. Bioeng.,* 69, 66 – 73.

In: Penicillin ISBN: 978-1-62100-233-8
Editors: G. Barresi and P. Komulainen

Chapter 3

NOVEL PEPTIDE ENDOWED WITH ANTIBIOTIC PROPERTIES

Séverine Padiolleau-lefevre[1,*], Denis Phichith[2,†] and Bérangère Avalle[1,‡]

[1]Génie Enzymatique et Cellulaire,
Université de Technologie de Compiègne, France
[2]Department of Physiology,
University of Pennsylvania School of Medicine, PA, US

ABSTRACT

Bacterial cell wall synthesis is a complex process that involves several key players. Among them, the Penicillin Binding Proteins (PBPs) and the ß-lactamases play a central role. The former are essential in cell wall synthesis. The latter appeared in response to decades of use of ß-lactam ring-based biomolecules acting as environmental selective pressures, and have become the most common efficient resistance systems developed by bacteria.

Our laboratory pioneered the generation of catalytic antibodies using the anti-idiotypic approach. Using this approach, we engineered a catalytic antibody displaying a ß-lactamase-like activity: 9G4H9. This antibody was shown to be the internal image of class A ß-lactamase, the

* E-mail: severine.padiolleau@utc.fr.
† E-mail: dphichit@mail.med.upenn.edu.
‡ E-mail: berangere.bihan@utc.fr.

starting antigen of the idiotypic cascade. By using this catalytic antibody as the target, we selected a cyclic heptapeptide (Pep90) from a phage-displayed library, and studied its biochemical properties. Not only does Pep90 bind the catalytic antibody, but it is able to competitively inhibit its catalytic activity through key residues.

Interestingly, Pep90 displayed striking inhibitory properties. Indeed, Pep90 was able to inhibit both TEM-1 ß-lactamase and some PBPs in a competitive manner. Few molecules are known to inhibiting both the antibiotic target (PBPs) and the resistance proteins (ß-lactamase). A molecule that displays both antibacterial and antibiotic properties, could potentially be used as a scaffold to generate peptidomimetic compounds. Indeed, the weakness of peptides as therapeutic agents is their tendency to undergo rapid proteolysis in serum. In order to overcome this limitation, we designed two peptidomimetics derived from Pep90 that contain unnatural amino_acids, and characterized their ability to inhibit both PBPs and ß-lactamase as peptidomimetic-like molecules. Our results suggest that Pep90 and its peptidomimetic counterparts may be good candidates for the development of new antibiotic molecules.

INTRODUCTION

As a collective idea, the problem of antibiotic resistance arose at the same time as penicillins were introduced in human therapeutics. While some strains developed long before the use of penicillins have been shown to harbour the genetic machinery of antibiotic resistance, the problem has grown to unprecedented proportions since the emergence of pathological strains able to resist a broad range of the antibiotic arsenal (Drawz and Bonomo, 2010). One of the major goals of therapeutic research is thus to curb the evolutionary impetus of such threatening strains by finding new molecules in a position to target the resistance mechanisms. Many leads may be prospected: inhibition of bacterial cell wall synthesis or of the ribosomal complexes, destabilization of the membrane… (Silver et al, 2010).

Among all the vectors of bacterial resistance, the ß-lactamases are hitherto the most effective. In response to decades of ß-lactam ring-based biomolecules acting as environmental selective pressures, these enzymes developed a highly efficient defence mechanism employed by bacteria. As trojan horses, they mimic the active site of the natural targets of penicillins, the Penicillin-Binding Proteins (PBPs), to capture the inhibitory molecules and destroy them rendering penicillins and their derivatives uneffective. The hypothesis whereby ß-lactamases and penicillin-binding proteins may have evolved from a common ancestor (Goffin and Ghuysen, 1998) is supported by (i) a conserved enzymatic mechanism involving an activated serine, and (ii) similar structural motifs, both enzymes adopting an α/β fold (Ghuysen, 1991; Galleni and Frère, 2007; Urbach *et al.* 2009).

The overall issue of antibiotherapy research is to design new sets of molecules, ones targeting PBPs and others targeting ß-lactamases, the panacea being molecules performing both functions. The spectrum of potentially reactive molecules is obviously very large, from organic molecules derived from penicillins to completely new biological structures. Considering that some derivatives might not, in the long-term, pass through the mesh of the bacterial resistance, biological structures appear to be the most promising. One way to obtain new molecules without a priori knowledge about the required structure is to use the power of combinatorial libraries. Biological supports are revolutionary tools from which a high diversity of molecules may be gained.

In this chapter, we present the pathway followed to obtain a new molecule endowed with antibiotic properties.

From the Enzyme to the Abzyme and Its scFv Counterpart

According to Jerne's hypothesis, the immune system is able to generate antibodies specific to variable regions of other antibodies (Jerne, 1974). Indeed, for each immunoglobulin (Ab1), a complementary antibody (Ab2) exists that is directed against the idiotypic determinant of Ab1. Some Ab2 molecules recognize idiotopes in the combining site of Ab1. Ab2 represents the internal image of the starting antigen, meaning that this former and Ab2 may carry the same properties (binding and/or catalysis). From this idea, we used ß-lactamase (EC 3.5.2.6) as a template to generate a catalytic antibody that displays ß-lactamase activity (Figure 1). The catalytic antibody 9G4H9 (Avalle *et al,* 1998) was generated using the internal properties of anti-idiotypic antibodies: 9G4H9 thus represents the internal image of the ß-lactamase active site. The catalytic mechanism of this abzyme was previously investigated (Avalle *et al,* 2000; Lefèvre *et al,* 2001, Padiolleau-Lefèvre *et al,* 2003). The single chain variable fragment (scFv) of 9G4H9 was cloned and produced in *Escherichia coli* (Padiolleau-Lefèvre *et al,* 2006), and displays similar kinetic features with the whole antibody counterpart, suggesting that scFv 9G4H9 should be a realistic study model of the abzyme. The kinetic

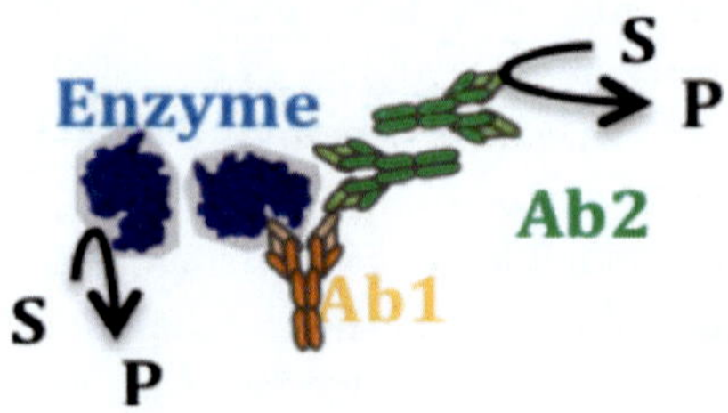

Figure 1. Schematic representation of idiotypic network manipulation in order to obtain catalytic antibodies.

characterization, associated to the known structure function relationship of ß-lactamase activity, led us to identify the residues involved in the active site, on the basis of a molecular model. These hypotheses were thus developed in accordance with ß-lactamase catalytic mechanistic and structural data (Meroueh *et al,* 2005; Fisher and Mobashery, 2009). The active site hypothesis of the 9G4H9 catalyst was finally validated by site-directed mutagenesis, demonstrating that Ser26 and Ser28 function as nucleophiles, and Glu98 as a general acid/base catalyst (Phichith *et al,* 2009).

Actually, this work led to the production of well known and characterized models of the ß-lactamase activity, constituting an interesting target for the selection of biomolecules acting as therapeutic agents.

THE PHAGE DISPLAY TECHNOLOGY: A POWERFUL TOOL TO SELECT ACTIVE BIOMOLECULES

Bacteriophages are viruses that exclusively infect bacterial cells, disrupting their metabolism. Mechanistically, bacteriophage viruses infect bacteria and use the bacterial machinery to replicate and produce new particles. In addition, bacteriophages can easily accommodate foreign DNA coding segments into their own genome. More specifically, insertion of DNA encoding the desired protein or peptide sequence downstream of *pIII* gene allows expression of the protein and presentation on the bacteriophage surface. The physical linkage between the interest gene and the displayed protein or peptide on the viral particle constitutes the key point of this technology. Phage display technology has been and still is a powerful tool over the past decade for protein design, evolution and drug development. The application fields are various: Phage display greatly benefited the antibody engineering field, from antibody affinity maturation (Thie *et al.*, 2009; Thompson *et al.*, 1996) and humanized antibody design (Rader *et al.*, 1998; Son *et al.*, 2004) to antibody-derived cancer drugs (Lamdan *et al.*, 2011; Yang *et al.*, 2007). Phage display has been used to improve enzymatic catalytic activities (Soumillion P. *et al.*, 1994) and convert enzymatic activities (Vichier-Guerre S. & Jestin J.L., 2010; Droge M.J *et al.,* 2006). Moreover, peptide display on phage has been successfully used to decipher binding motifs (Koivunen E. *et al.*, 1995) and develop biosensors (Wu J. *et al.*, 2010) and biomarkers (Park J.P. *et al.*, 2010). In summary, phage-display has become a powerful tool for protein evolution, motif design, and drug screening over the past decade.

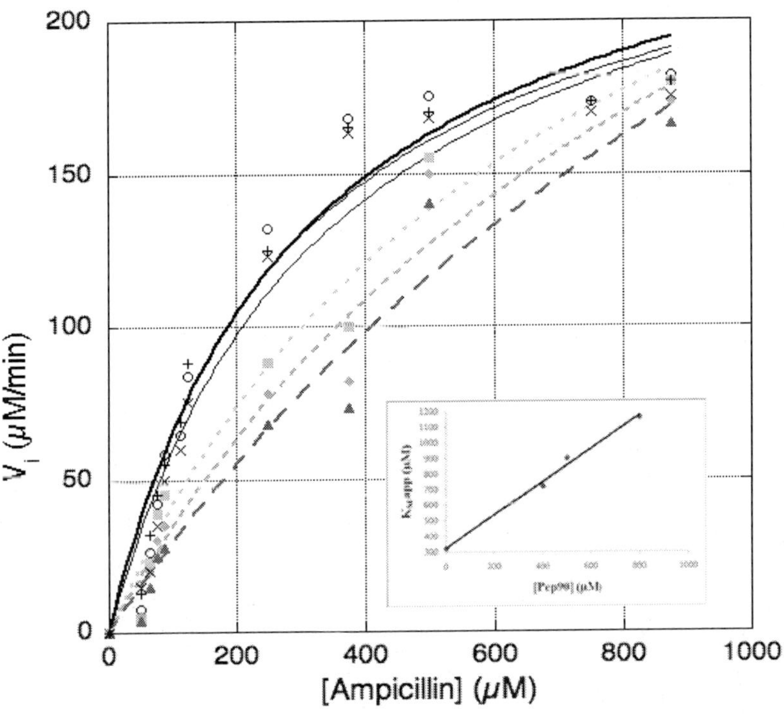

Figure 2. Kinetics of ampicillin hydrolysis by ß-lactamase TEM-1 in the presence of Pep90.

(A): ○: without Pep90. □: [Pep90] = 200 µM. ◇ : [Pep90] = 400µM. △: [Pep90] = 800µM.

(B): Apparent K_M following of the reaction as a function of the concentration of Pep90.

SELECTION OF A PEPTIDE INHIBITING A ß-LACTAMASE LIKE ACTIVITY, AND CHARACTERIZATION OF ITS INHIBITORY PROPERTIES

A library of random peptide sequences was used to select peptides that inhibit the anti-idiotypic catalytic antibody 9G4H9. This library displays cyclic peptides on the surface of bacteriophages. Based on various selection steps, including both the enrichment of peptides that bind the antibody and the discrimination of peptides that bind to part of the molecule other than the active site, the process allowed us to select a promising peptide, named Pep90, able to selectively inhibit the IgG 9G4H9 (Yribarren *et al,* 2003). Pep90 is a cyclic peptide flanked by two cysteine residues at its N and C termini. It has a loop structure similar to the BLIP loop that is involved in TEM-1 β-lactamase inhibition (Rudgers *et al.* 1999; Rudgers *et al.* 2001).

We investigated the inhibitory capabilities of Pep90 by measurement of the residual ß-lactamase activity of 9G4H9 after 20 min of incubation with this inhibitor. The half inhibitory concentration (IC_{50}) was thus estimated around 9.10^{-5}M. Experiments of surface plasmonic resonance were performed and revealed a high affinity of Pep90 for IgG 9G4H9, with a K_D value of 7.4×10^{-8}M (table 1). Considering its inhibition properties toward 9G4H9, we measured the effect of Pep90 on ß-lactamases (Figure 2).

As shown in Figure 2, Pep90 behaves like a_competitive inhibitor of ß-lactamase TEM-1. Pep90 inhibits TEM-1 ß-lactamase with a K_i within the micromolar range: $K_i = 333 \pm 13$ μM. Furthermore, we measured no inhibitory effect of the linear counterpart of Pep90. Cyclic pep90 has better inhibitory properties than linear pep90 because its cyclic structure minimizes steric hindrance and thus, enhances its inhibitory properties.

Thus, Pep90 inhibits not only 9G4H9, but also the TEM-1 ß-lactamase. We previously demonstrated that 9G4H9 shows features that are a hybrid of the ß-lactamase and PBPs, two related bacterial enzymes whose genes divergently evolved from the same ancestral gene. Since Pep90 inhibits both 9G4H9 and ß-lactamase, it may perform a cross-competitive_inhibition of some PBPs. We measured IC_{50} of several PBPs (Phichith *et al,* 2010), and also determined binding constants by surface plasmon resonance using biotinylated Pep90 as ligand and various ß-lactamases and PBPs as analytes (Table 1). The results demonstrated that Pep90 tightly interacts with ß-lactamases and PBP active sites that are closely related to the model antigen used to produce IgG 9G4H9. We further characterized Pep90 at the amino-acid level.

Table 1. Affinity of Pep90 for various catalysts

Catalysts		K_D (µM)
Abzyme	IgG 9G4H9	0.074
	scFv 9G4H9	5.2
Class A ß-lactamases	TEM1 - E. *coli*	15.5
	TEM1 - B. *cereus*	152
	BS3 TEM1 - B. *licheniformis*	-
Class C ß-lactamase	P99 - E. *cloacae*	-
Class D ß-lactamase	OXA-10 - P. *aeruginosa*	-
PBPs	K15 DD-transpeptidase - Streptomyces	-
	PBP5 DD-carboxypeptidase - E. *coli*	1.9
	R61 DD-carboxypeptidase, transpeptidase - Streptomyces	3.8

Surface Plasmon Resonance (SPR) analysis was performed with a BIAcore 3000 (BIAcore GE, USA). Sensor Chip SA (Streptavidin chip) was purchased from BIAcore. Immobilization of the biotinylated Pep90 ligand on the surface of SA chip was done by injecting 50 µl of a solution containing 0.1µM biotinylated Pep90 at a flow rate 7 µL/minute. The K_D values were determined by injecting 40 µl of catalyst and penicillin binding proteins solution ranging from 0.5 to 200 µM at a flow rate of 10 µL/minute. The regeneration of the surface was carried out by injecting 6 µl of 50 mM HCl. Curve fitting was carried out using BIAevaluation 3.0.1 software (BIAcore GE, USA), and based on a Langmuir model.

Highlight of Amino Acids from Pep90 that Interact with TEM-1 ß-lactamase Active Site

Since Pep90 is displayed on the M13 bacteriophage, it is easy not only to access its sequence (CYHFLWGPC), but also to perform an “alanine-scanning” by DNA mutagenesis to generate Pep90 variants, and thus to determine which residues of Pep90 are likely to interact with the ß-lactamase active site. We further performed a phage-ELISA test in order to determine which residues of Pep90 are likely to interact with the ß-lactamase active site. For this purpose, and on the basis of the known BLIP residues implied in

A

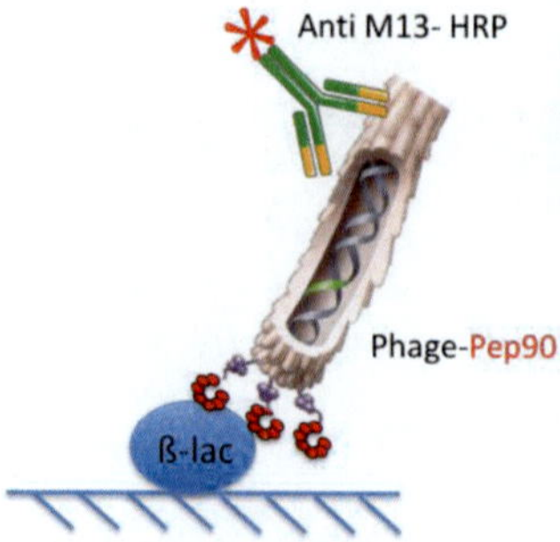

B

Peptide	Binding
Pep Ctrl	-
Pep90 *Wild type*	+
Pep90 Y/A	+
Pep90 F/A	+
Pep90 W/A	-
Pep90 P/G	-
Pep90 G/P	+

Figure 3. Principle of the phage-ELISA system and qualitative effect of Pep-90 variants on recognition of ß-lactamase.

(A): Scheme of the Phage-ELISA capture format.
(B): Qualitative effect of Pep90 derivatives on ß-lactamase from *Escherichia. coli*. The binding of Pep90 and its derivatives was evaluated by phage ELISA. (+) indicates an absorbance value at least twice greater than the signal obtained with the negative control (Phage-Pep95). (-) corresponds to signals identical to the negative control.

interaction with the enzyme (Rudgers *et al.* 1999; Rudgers *et al.* 2001), we alternately mutated the aromatic residues of Pep90, i.e. Tyr, Phe, and Trp. We also converted the G to P and vice-versa, in order to obtain more accurate information on the role of the peptide flexibility in the recognition process. A phage-ELISA of the phage-Pep90 variants on TEM-1 ß-lactamase was also carried out (Figure 3). From qualitative measurements we can conclude, on the one hand, that tryptophan is mainly important for the interaction of Pep90 with its target, and on the other hand, that the presence of proline ensures the necessary rigidity of the peptide. The inhibitory effect of derivatives of Pep90 was then measured on the catalytic activity of ß-lactamase TEM-1 (Phichith *et al,* 2010). These kinetic studies clearly indicate that the central tryptophan is essential for the inhibitory effect of Pep90 on ß-lactamase.

A structural approach was then pursued. On the basis of free open source software (SPDMOD, Kong *et al,* 2004), a structural model of Pep90 was proposed (Phichith *et al*, 2010) (Figure 4A). This conformation respects the phi and psi angles, and illustrates the accessibility of the central tryptophan. We further focused on the Pep90/TEM1 ß-lactamase interaction model. By using the GRAMM-X program (Tovchigrechko and Vakser, 2006), we constructed a Pep90/ TEM-1 ß-lactamase complex model. By indicating the interacting residues, we obtained a refinement of the docking model. Several acceptable models were generated in accordance with the energy minima.

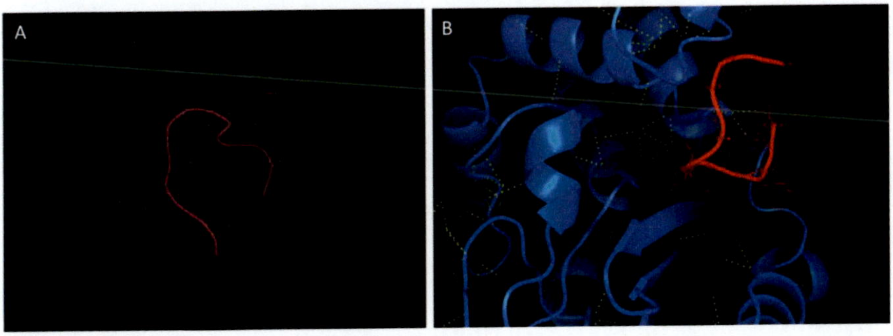

Figure 4. Molecular modelling of the cyclic heptapeptide Pep90 and the docking result between TEM-1 ß-lactamase and its inhibitor Pep90.

(A) The 3D Pep90 model was generated by SPDMOD simulation, using the streptavidin cyclic ligand as matrix.
(B) TEM-1 ß-lactamase /Pep90 docking model. TEM-1 ß-lactamase is illustrated in blue and the Pep90 in red.

Among them, a significant number (30%) showed that Pep90 specifically interacts with the active site of TEM-1 ß-lactamase (Figure 4B). In all cases, the disulfide bridge of Pep90 that allows its cyclization is oriented towards the exterior of the active site.

TOWARDS THE USE OF PEPTIDOMIMETICS

As a consequence of the extensive use of antibiotics, bacteria have developed highly effective defence systems. The development of new molecules with new mechanisms of action, appear is now essential to fight against microbial infections (Bragonzi, 2010). Natural peptides are small biomolecules that have recently generated great interest in drug design. Their major drawbacks include sensitivity to proteolytic degradation under physiological conditions and side effects when bound to several types of receptors. In order to circumvent these features, peptidomimetics provide an interesting alternative to natural peptides (Vaara, 2011). There is growing interest in peptidomimetics due to their greater resistance to proteases, higher lifetime activity, better cellular distribution and fewer side effects. In this context, naturally occurring peptides make interesting starting points for the design and synthesis of biologically active peptidomimetics, as illustrated by recent works of Srinivas and collaborators (2010). Pep90 appears to be particularly a good candidate for a new peptidomimetic scaffold due to its ability to inhibit both TEM-1 β-lactamase (K_i = 333 μM) and several penicillin-binding proteins from different bacteria species with IC_{50} ranging from 6 to 100 μM. Based on our results, showing the inhibition properties of Pep90 were mainly due to W and P residues, two peptidomimetics were synthesized (Eurogentec, Belgium): Pep90-W_{Aza} and Pep90-P_D

Pep90-W_{Aza}: CYHFL$W_{D\text{-}L\text{-}7\text{-}azatropine}$GPC
Pep90-P_D: CYHFLWP_D

Incubation of ß-lactamase with various concentrations of Pep90-W_{AZA} and Pep90-P_D ranging from 0 to 400 μM showed the maintenance of inhibitory ability of both peptidomimetics (Phichith *el al*, 2010). When comparing the IC_{50} value of Pep90 with its peptidomimetics counterparts (table 2), we conclude that despite slightly higher values for the peptidomimetics, the inhibition is nevertheless maintained when Trp and Pro are replaced by unnatural amino acids.

Table 2. Inhibitory properties of Pep90 and peptidomimetics counterpart toward the TEM1 ß-lactamase

Inhibitor	IC_{50} (μM)
Pep90 wild type	333.0 ± 13.0
Pep90-W_{AZA}	592.2 ± 29.6
Pep90-P_D	436.3 ± 21.8
Pep95	-

The Pep90 peptidomimetics counterpart Pep90-W_{AZA} and Pep90-P_D were made to order by Eurogentec (Liège, Belgium). They include either a tryptophan analog (dl-7-azatryptophan, WAza) or enantiomer D of proline (P_D). TEM-1 ß-lactamase at 3.5 nm was pre-incubated at room temperature for 45 min in 100 mm phosphate buffer, pH 7.4, supplemented with 2.4% dimethylsulfoxide and 0.01% gelatine, with various concentrations of peptidomimetics ranging from 0 to 400 μm. Ampicillin hydrolysis at 220 μm final concentration was followed spectrophotometrically at 240 nm. ß-lactamase residual activity was normalized to that for ampicillin hydrolysis without peptide.

CONCLUSION

Pep90 selected from antibody 9G4H9 inhibits the activity of the anti-idiotypic antibody. It also inhibits ß-lactamase itself. Among aromatic residues, only tryptophan is necessary to maintain the inhibitory properties of Pep90. The presence of tryptophan is necessary but not sufficient for the inhibitory activity: we showed that when the enzyme is saturated by soluble tryptophan (maximum concentration= 9 mM at room temperature, for 45 minutes) no inhibition is measured (data not shown). The accurate orientation of tryptophan is required for its positioning within the catalytic cavity of ß-lactamase.

The rigidity of the molecule, induced by both the cyclization and the presence of a terminal proline, also appears to be an important criterion. Furthermore, Pep90 is not only able to inhibit scFv 9G4H9 and TEM-1 ß-lactamase of *Escherichia Coli* but also several low and high molecular weight penicillin-binding proteins.

The main goal in small biomolecule-based drug development is to modulate wild-type protein activities. Small biomolecules are the most promising path to new drug discovery and optimization (Kieber-Emmons et *al.*, 1997). Recent progress in drug design has opened a new field of application for potential peptidomimetic-based ligands, and biotech companies are increasingly interested in designing new peptidomimetic inhibitors as lead drugs. In response to decades of environmental selective pressure from β-lactam ring based biomolecules, bacteria have become multi-drug resistant to most common known antibiotics. In order to circumvent this multiresistance, new antibiotic drugs must be developed. This is an ongoing challenge for public health. In this context, Pep90 appears to be a good candidate, since it combines two functions: on the one hand, it is able to inhibit enzymes involved in the cell wall synthesis, and in the other hand it is able to clamp one of the major bacterial resistance mechanisms.

REFERENCES

Avalle, B., Débat, H., Friboulet, A. and Thomas, D. (2000). Catalytic mechanism of an abzyme isplaying a ß-lactamase-like activity. *Appl. Biochem. Biotechnol. 83,* 163-171.

Avalle, B., Thomas, D., and Friboulet, A. (1998). Functional mimicry: elicitation of a monoclonal anti-idiotypic antibody hydrolysing ß-lactams. *FASEB J. 11,* 1055-1060.

Bragonzi A. (2010). Fighting back: peptidomimetics as a new weapon in the battle against antibiotic resistance. *Sci. Transl. Med. 2,* 21ps9.

Conrad, U., and Scheller, J. (2005). Considerations on antibody-phage display technology. *Comb. Chem. High Throughput Screen. 8*, 117-126.

Drawz, S.M., and Bonomo, R.A. (2010). Three decades of beta-lactamase inhibitors. *Clin. Microbiol. Rev. 23,* 160-201.

Droge M.J., Boersma Y.L., Van Pouderoyen G., Vrenken T.E., Ruggeberg C.J., Reetz M.T., Dijkstra B.W., and Quax W.J. (2006) Directed evolution of Bacillus subtilis lipase A by use of enantiomeric phosphonate inhibitors: crystal structures and phage display selection. *Chembiochem. 1,* 149-157.

Fisher, JF. and Mobashery, S. (2009). Three decaes of the class A ß-lactamase acyl-enzyme. *Curr. Protein Pept. Sci. 10,* 401-407.

Galleni, M. and Frère, J.M. (2007). Kinetics of beta-lactamases an penicillin-binding proteins, p195-213. In R.A. Bonomo and M.E Tomasky (ed.), Enzyme-mediated resistance to antibiotics: mechanisms, dissemination, and propects for inhibition. ASM Press, Washington DC.

Ghuysen, J.M. (1991). Serine beta-lactamases and penicillin-binding proteins. *Annu. Rev. Microbio. 250,* 313-324.

Goffin C. and Ghuysen JM. (1998). Multimodular penicillin-binding proteins: an enigmatic family of orthologs and paralogs. *Microbiol. Mol. Biol. Rev. 62,* 1079-1093.

Jerne N. (1974). Toward a network theory of the immune system. *Ann. Immunol. (Paris). 125c,* 373-389.

Kieber-Emmons T., Murali R. and Greene MI. (1997) Therapeutic peptides an peptidomimetics. *Curr. Opin. Chem. Biol. 8,* 435-441.

Koivunen E., Wang B., and Ruoslahti E. (1995) Phage libraries displaying cyclic peptides with different ring sizes: ligand specificities of the RGD-directed integrins. *Biotechnology. 13,* 265-270.

Kong L., Lee K., Tong JC., Tan TW., and Ranganathan S. (2004) SDPMOD: an automated comparative modelling server for small disulphide-bonded proteins. *Nucleic Acids Res. 32,* (WebServer issue): W356-W359.

Lamdan H., Ayala M., Rojas G., Munoz Y., Morera Y., Guirola O., Chinea G., and Gavilondo J.V. (2001) Isolation of a novel neutralizing antibody fragment against human vascular endothelial growth factor from a phage-displayed human antibody repertoire using an epitope disturbing strategy. *J. Biotechnol. 151,* 166-174.

Lefèvre, S., Débat, H., Thomas, D., Friboulet A. and Avalle, B. (2001). A suicide substrate mechanism for hydrolysis of ß-lactams by an anti-idiotypic catalytic antibody. *FEBS Lett. 489*, 25-28.

Meroueh SO., Fisher JF., Schlegel HB., and Mobashery S. (2005). Ab inition QM/MM study of class A ß-lactamase acylation: dual participation of Glu166 and Lys73 in a conserted base promotion of Ser70. *J. Am. Chem. Soc. 127,* 15397-15407.

Padiolleau-Lefèvre S., Débat, H., Phichith, D., Thomas, D., Friboulet, A., and Avalle, B. (2006). Expression of a functional scFv fragment of an anti-idiotypic antibody with a β-lactam hydrolytic activity. *Immunol.. Lett. 103,* 39-44.

Padiolleau-Lefèvre, S., Débat, H., Thomas, D., Friboulet A. and Avalle, B. (2003). *In vivo* évolution of a ß-lactamase-like activity throughout the idiotypic pathway. *Biocatal. Biotransfor. 21,* 79-85.

Park J.P., Cropek D.M., and Banta S. (2010) High affinity peptides for the recognition of the heart disease biomarker troponin I identified using phage display. *Biotechnol. Bioeng. 105,* 678-686.

Phichith D., Bun S., Padiolleau-Lefèvre S., Banh S., Thomas D., Friboulet A., and Avalle B. (2009). Mutational and inhibitory analysis of a catalytic antibody. Implication for drug discovery. *Mol. Immunol. 47,* 348-356.

Phichith D., Bun S., Padiolleau-Lefèvre S., Guellier A., Banh S., Galleni M., Frère JM., Thomas D., Friboulet A., and Avalle B. (2010). Novel peptide inhibiting both TEM-1 ß-lactamase and penicillin binding proteins. *FEBS J. 277,* 4965-4972.

Rader C., Cheresh D.A., and Barbas C.F 3rd. (1998). A phage display approach for rapid antibody humanization: designed combinatorial V gene libraries. *Proc. Natl. Acad. Sci. USA 95*, 8910-8915.

Rudgers GW., Huang W., and Palzkill T. (2001). Binding properties of a peptide derived from ß-lactamase inhibitory protein. Antimicrob. *Agents Chemoth. 45*, 3279-3286.

Rudgers GW. and Palzkill T. (1999). Identification of residues in ß-lactamase critical for binding ß-lactamase inhibitory protein. *Antimicrob. Agents Chemoth., 274,* 6963- 6971.

Sedlacek, R. and Chen, E. (2005). Screening for protease substrate by polyvalent phage display. *Comb. Chem. High Throughput Screen. 8*, 197-203.

Sidhu, S.S., and Koide, S. (2007). Phage display for engineering and analyzing protein interaction interfaces. *Curr. Opin. Struct. Biol. 17*, 481-487.

Silver, L.L. (2010). Challenges of antibacterial discovery. *Clin. Microbiol. Rev. 24*, 71-109.

Simon, M.D., Feldman, M.E., Rauh, D., Maris, A.E., Wemmer, D.E., and Shokat, K.M. (2006). Structure and properties of a re-engineered homeodomain protein-DNA interface. *ACS Chem. Biol. 1*, 755-760.

Son J.H., Lee J.J., Kwon B., Kwon B.S., and Park J.W. (2004) Humanization of agonistic anti-human 4-1BB monoclonal antibody using a phage-displayed combinatorial library. *J. Immunol. Methods. 286,* 187-201.

Soumillion P., Jespers L., Bouchet M., Marchand-Brynaert J., Winter G., and Fastrez J. J. (1994) Selection of beta-lactamase on filamentous bacteriophage by catalytic activity. *Mol. Biol. 237,* 415-422.

Srinivas N., Jetter P., Ueberbacher BJ., Werneburg M., Zerbe K., Steinmann J., Van der Meijden B., Bernardini F., Lederer A., Dias RL., Misson PE., Henze H., Zumbrunn J., Gombert FO., Obrecht D., Hunziker P., Schauer S., Ziegler U., Käch A., Eberl L., Riedel K., DeMarco SJ., and Robinson JA. (2010) Peptidomimetic antibiotics target *Science 327,* 1010-1013.

Thie H., Voedisch B., Dubel S., Hust M., and Schirrmann T. (2009) Affinity maturation by phage display. *Methods Mol. Biol. 525,* 309-322.

Thompson J., Pope T., Tung J.S., Chan C., Hollis G., Mark G., and Johnson K.S. (1996) Affinity maturation of a high-affinity human monoclonal antibody against the third hypervariable loop of human immunodeficiency virus: use of phage display to improve affinity and broaden strain reactivity. *J. Mol. Biol. 256,* 77-88.

Tovchigrechko A. and Vakser IA. (2006) GRAMM-X public web server for protein-protein docking. *Nucleic Acids Res. 34*, 310-314.

Urbach, C., Evrard, C., Pudzaitis, V., Fastrez, J., Soumillion, P. and Declercq, J.P. (2009): Structure of PBP-A Thermosynechococcus elongatus, a penicillin-binding protein closely related to class A beta-lactamases. *J. Mol. Biol***.** *386,* 109-120.

Vaara M. (2009). New approaches in peptide antibiotics. *Curr. Opin. Pharma. 9,* 571-576.

Vichier-Guerre S. and Jestin J.L. (2010). Characterisation of a DNA polymerase highly mutated along the template binding interface. *Mol. Biotechnol. 46,* 58-62.

Wu J., Cropek D.M., West A.C., and Banta S. (2010). Development of a troponin I biosensor using a peptide obtained through phage display. *Anal Chem. 82,* 8235-8243.

Yang G.H., Yoon S.O., Jang M.H., and Hong H.J. (2007). Affinity maturation of an anti-hepatitis B virus PreS1 humanized antibody by phage display. *J. Microbiol. 45,* 528-533.

Yribarren, A.S., Thomas, D., Friboulet A. and Avalle, B. (2003). Selection of peptides inhibiting a ß-lactamase-like activity. *Eur. J. Biochem. 270,* 2789–2795.

In: Penicillin
Editors: G. Barresi and P. Komulainen
ISBN: 978-1-62100-233-8

Chapter 4

EFFECTS OF PENICILLIN ON CLINICAL TESTS WITH SPECIAL REFERENCE TO SERUM ALBUMIN MEASURED BY A MODIFIED BCP METHOD

Yuzo Kayamori, Miyuki Ono and Dongchon Kang *
Department of Clinical Chemistry and Laboratory Medicine,
Kyushu University Graduate School of Medical Sciences, Japan

ABSTRACT

Penicillin derivatives are widely used for treatment of bacterial infection. A high-dose penicillin is recommended as a first choice particularly for infective endocarditis in the guideline from the American Heart Association. We found two patients with infective endocarditis who had serum albumin shifting to a cathode side on the electrophoresis. The both patients had been treated with a high-dose penicillin G. The serum albumin concentrations of the patients were 1.7 g/dL and 1.8 g/dL when measured by a modified bromocresol purple (BCP) method while they were 3.4 and 3.3, respectively, by a bromocresol green (BCG) method. To confirm these low values obtained by the modified BCP method are

* Correspondence to Dongchon Kang, Department of Clinical Chemistry and Laboratory Medicine, Kyushu University Graduate School of Medical Sciences, 3-1-1 Maidashi, Higashi-ku, Fukuoka 812-8582, Japan. Fax +81-92-642-5772; E-mail kang@mailserver.med.kyushu-u.ac.jp.

due to the high-dose penicillin G therapy, we added penicillin G in normal sera and measured albumin concentration by the BCP and BCG methods. Exogenously added penicillin G inhibited the BCP reaction in dose-dependent and incubation time-dependent manners but not the BCG reaction. Here, we generally describe effects of penicillin derivatives on serum albumin measurement and other clinical tests.

INTRODUCTION

The mechanism by which drugs influence clinical laboratory test value is roughly divided into two patterns. One is results from their own pharmacologic effects on tissues in vivo, either desired or undesired: for example, elevation of serum AST due to liver toxicity. The other one is that a drug and/or its metabolites interfere directly with assay reaction(s). In this chapter, we describe on the latter because the pharmacologic action and the side effect of penicillin are described elsewhere in this book.

There have been some reports that penicillin or its derivatives chemically interfere with measurement systems of clinical laboratory test and consequently cause abnormal test values. The most cases are from interactions of these drugs with serum protein, especially albumin.

Serum albumin is a protein with a molecular weight of about 66 kDa and synthesized by liver. The albumin occupies 50-65% of serum protein. Serum albumin plays many functions in blood such as pH buffering, maintenance of the osmotic pressure, a source of amino acids for peripheral tissues, and antioxidative effects. In addition, it has a large binding capacity of endogenous and exogenous materials such as bilirubin, mineral ions, and administered drugs. Low serum albumin occurs in many cases, e.g. malignancy, liver disease, nephrotic syndrome, burns, protein-losing enteropathy, malabsorption, and malnutrition [1]. Therefore, serum albumin is one of the most frequently and routinely measured clinical tests

Currently, serum or plasma albumin is measured mainly by two methods, bromocresol green (BCG) and bromocresol purple (BCP) methods. However, the BCG method has a problem with specificity for albumin because the BCG dye reacts with various globulins as well as albumin [2]. On the other hand, the BCP dye was reported to be more albumin-specific than BCG [3]. Nonetheless, initially BCP was hardly used because its reactivity with albumin is strongly affected by redox states of thiol (SH) groups of an albumin molecule. A new reagent that improves the problem of BCP is developed and then the modified BCP using it has become used widely for the clinical laboratory test [4]. According to recent College of American Pathologists (CAP) Comprehensive Chemistry report [5], the BCG method and the BCP method occupies predominant shares, 47% and 52%, respectively.

Here, we introduce two cases with infective endocarditis whose serum albumin values measured by the BCP method were much lower than those anticipated by electrophoretic evaluation [6]. We found that penicillin administered to the patients was a cause for the underestimation.

INFLUENCE ON SERUM ALBUMIN VALUE OF INFECTIVE ENDOCARDITIS PATIENT

The profile of the two patients with infective endocarditis was shown in Table 1. The clue started from a finding that the patient serum albumin was electrophoresed broadly and shifted to a cathode on agarose gel electrophoresis compared with the normal serum as shown in Figure 1. Then we noticed the discrepancy of their serum albumin values between the BCP and electrophoretic methods. A high dose of antibiotics are usually administered to patients with infective endocarditis. In the guideline of prevention and the diagnosis of the infective endocarditis, penicillin is a first choice as an antibacterial agent [7]. In fact, the two patients were treated with a high dose of penicillin G.

Penicillin is already reported by Arvane et al. to cause abnormal electrophoresis of serum albumin [8]. Various inherited albumin variants show abnormal electrphoretic patterns called bisalbuminemia. Arvane et al. have found that serum albumin of patients treated intravenously with penicillin show a similar electrophoretic pattern to the bisalbuminemia and this abnormal pattern disappears after the penicillin administration is discontinued. The degree of the abnormal electrophoresis was related to the penicillin concentration in serum and so they conclude the abnormal component is the penicillin-albumin complex. They describe that penicillin does not affect serum albumin measurement by a salt precipitation or 2-(4-hydroxybenzeneazo) benzoic acid (HABA) dye-binding method.

Table 1 shows that the serum albumin measured by the BCP method is much lower than those by the BCG and electrophoresis methods. Taken together with the results by Arvane et al., particularly only the BCP method seems strongly affected by penicillin.

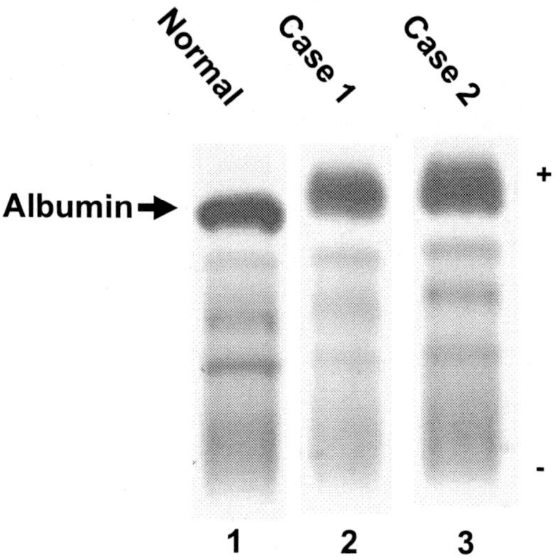

Figure 1. Assessment of serum albumin by electrophoresis. The sera were electrophoresed and stained with Ponso S. Lane 1, healthy control; lanes 2 and 3, cases showing abnormal albumin values. Symbols of + and – mean cathode and anode sides, respectively.

Table 1. Profile of laboratory data

	Case 1	Case 2
TP g/dL	6.3	6.3
ALB(modified BCP method) g/dL	1.7	1.8
ALB(BCG method) g/dL	3.4	3.3
serum protein electrophoresis		
ALB %	56.7	49.6
ALB g/dL (calculated by serum protein electrophoresis)	3.6	3.3

Serum total protein and albumin was measured with a P module 7700 (Hitachi High Technologies, Tokyo Japan) using Aqua Auto KainosTP-II (Kainos, Tokyo Japan) for total protein, Laboseed II ALB (Sino-test, Tokyo Japan) and Aqua Auto Kainos for the BCG and the modified BCP methods, respectively. Serum total protein was quantified by a biuret method. Human serum (0.4 μl) was electrophoresed on a SEPARAX-SP sheet (Fuji film, Tokyo) in barbital buffer C (Olympus, Tokyo) using an auto-electrophoretic system AES630 (Olympus). The proteins were stained with Ponso S (Olympus).

EFFECTS OF PENICILLIN G ON ALBUMIN MEASUREMENT

Both of the two cases were patients with infective endocarditis and treated with 8 ~24 million units of penicillin G every day for about one month. To examine if penicillin G indeed affects the albumin measurement by the BCP method, we directly added various amounts of penicillin G to normal serum. The serum was incubated for 12 h at 37°C and then electrophoresed. The albumin band was broaden and shifted to a cathode direction in a dose dependent manner of penicillin G (Figure 2). Next, we quantified albumin in the penicillin G-added serum. Penicillin G lowered the apparent albumin values in a dose-dependent manner when measured by the modified BCP method (Figure 3). A nearly maximal effect was observed at 10,000 units/mL. On the other hand, the BCG method was not affected at all (Figure 3). The degree of the electrophoretic change of the albumin band almost parallels with the apparent decline of albumin values measured by the modified BCP method (compare Figures 2 and 3). The maximal inhibitory effect of penicillin G on the BCP measurement was seen at 12 h-incubation while the BCG method was not affected even at 36 h-incubation (Figure 4). These results strongly suggest that the observed abnormal albumin values in the two patients were caused by the administrated large amount of penicillin G for treatment.

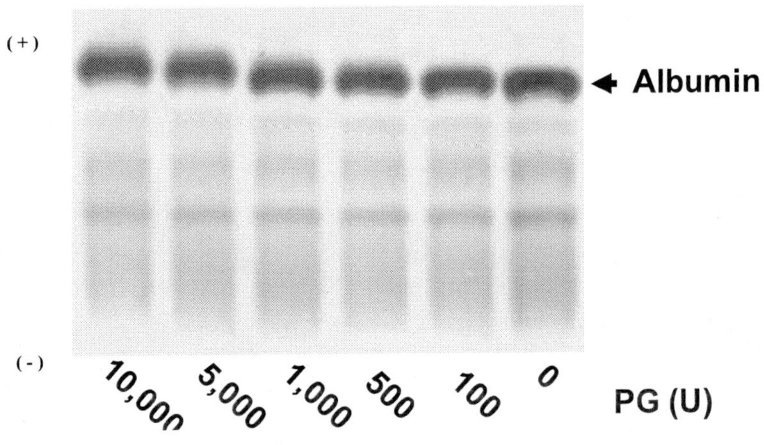

Figure 2. Effects of Penicillin G. Penicillin G was added to normal serum at indicated concentrations. After the sera were incubated for 12 h at 37°C, albumin concentrations were measured by the BCG and modified BCP methods. Each point is a mean of duplicated samples.

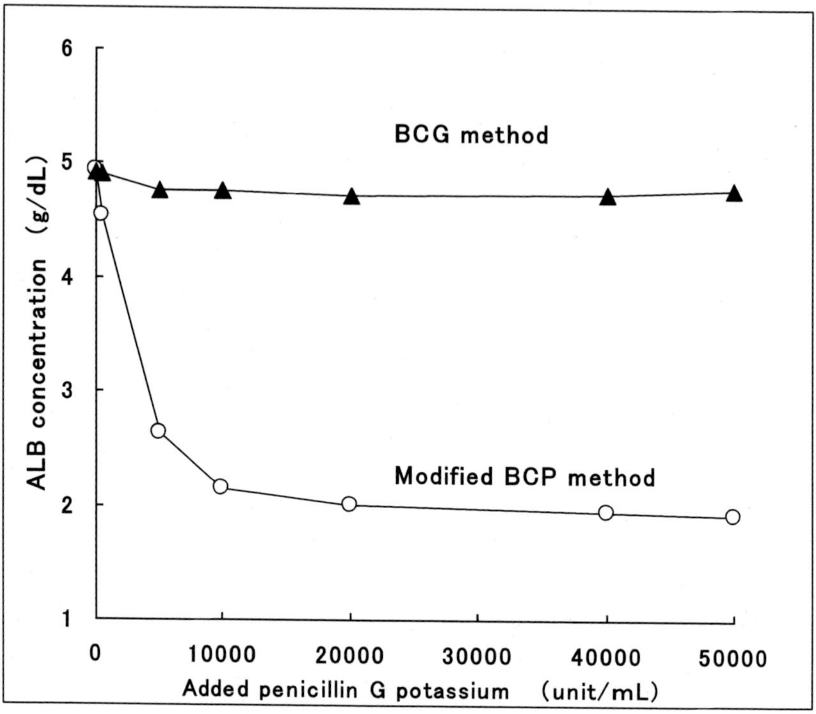

Figure 3. Effects of Penicillin G on electrophoresis. Penicillin G was added to normal serum at indicated concentrations. The sera were incubated for 12 h at 37°C and then electrophoresed as in Figure 1. PG, Penicillin G.

Influence of Other Antibiotics

Here, 14 drugs shown in Table 2 were examined, three of them are described in the guideline of prevention and the diagnosis of the infective endocarditis [7]. Each antibiotic was added to normal serum as shown in Table2. After 4-h incubation at room temperature, albumin concentration of each sample was determined by the BCG and BCP methods. These concentrations examined are around 10 to 100 times of the anticipated maximum therapeutic blood concentrations as shown in Table 2. In spite of extremely high doses used, the serum albumin measurements by the BCP or BCG methods were little affected with these antibiotics. Thus, the BCP method seems specifically affected by penicillin G.

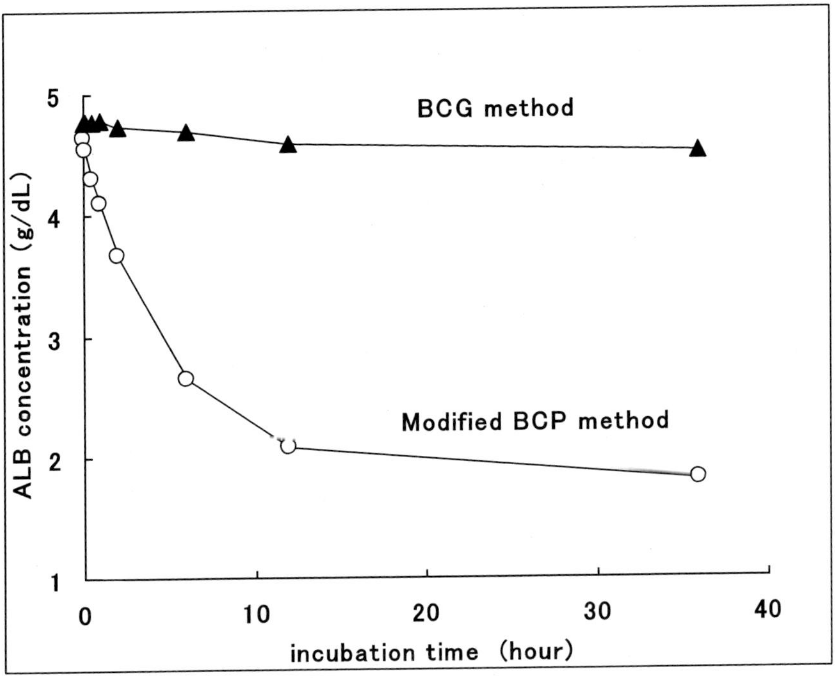

Figure 4. Time course of Penicillin G Effects. Penicillin G was added to normal serum at 10,000 units/mL. The serum was incubated at 37°C. At indicated time, an aliquot was taken and its albumin concentration was measured by the BCG and modified BCP methods. Each point is a mean of duplicated samples.

PROPERTIES OF THE BCG AND BCP METHODS

Albumin is an acidic protein with isoelectroric point (PI) 4.2-4.8 and can bind many endogenous as well as exogenous substances. Using this property, dye-binding methods are widely used nowadays like the BCG and BCP methods for quantification of serum albumin (Figure 5). Irrespective of the two methods are commonly based on the same principle, only BCP is affected by penicillin G. Therefore, we describe characters of the two dyes, which may be important to understand the interference mechanism.

Table 2. Effects of various antibiotics on serum albumin measurement

	Antibiotics				Effect of drugs		
	Class	Subgroup	Members	Maximal Concent-ration (mg/dL)*	Concentration in serum (mg/dL)	% of control	
						BCP method	BCG method
1					200	99	100
					3000	95	97
2	β-Lactam	cephems	cefamandole	17.3	200	97	100
3	β-Lactam	cephems	cephalexin	4.1	200	99	101
4	β-Lactam	cephems	cephalo-sporin-C		200	98	100
5	β-Lactam	cephems	cephalothin Na	8.8	200	97	100
6	β-Lactam	cephems	cefuroxime	9.92	200	98	99
7	β-Lactam	cephems	cefotaxime	23.83	200	99	100
8	β-Lactam	cephems	cefoxitin	25.2	200	98	100
9	β-Lactam	penicillin	ampicillin	3.5	200	99	101
					3000	97	98
10	β-Lactam	penicillin	penicilin G	300	600	88	99
					3000	61	97
11	Macrolide	14-mem-bered lactone	erythromycin	0.204	200	103	103
12	Macrolide		streptomycin	4	200	99	100
13	Tetracycline		tetracycline	0.28	200	99	97
14	Polypeptide		polymyxin		200	99	101
15	Others		chloramphenicol	0.75	200	103	103

* Usual therapeutic maximum concentration in blood described in "Collection of Medicine Japan Pharmaceutical Products ninth edition".

1. Bromcresol Green (BCG) Method

The BCG (bromcresol green, 3, 3', 5, 5'-tetrabromo-*m*-cresolsulfonphtalein) binds to serum albumin through its sulfophthalein and develops bluish green color. Since initial report by Rodkey et al. in 1965 [9], the BCG method has been improved in many points and is currently widely used in an automated manner. Albumin reacts fast with BCG within one minute while α- and β- globulins including acute phase reactants slowly react

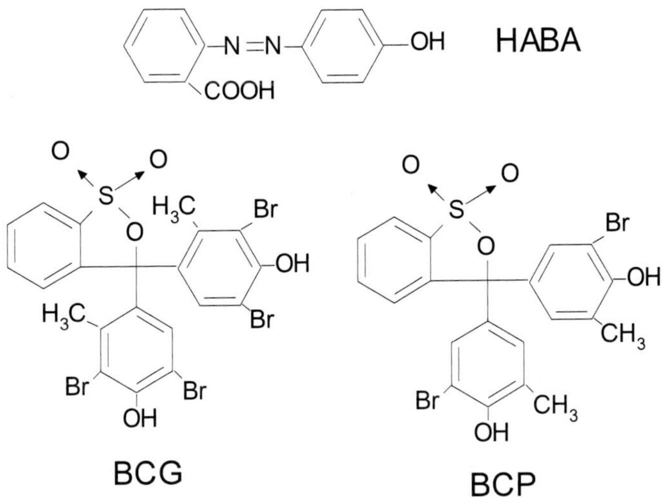

Figure 5. Chemical structure of albumin-binding dyes. HABA, 2-4-hydroxybenzeneazo benzoic acid; BCG, bromcresol green; BCP, bromcresol purple.

with BCG (taking 10-30 min). To minimize the nonspecific binding to globulins, most automated analyzers strictly control the measurement time and succeed in providing results comparable to immunoelectrophoresis [10-11]. However, Murakami et al. reported the reaction of the BCG-dye reacts with the globulin significantly even within 15 s [12].

2. Bromcresol Purple (BCP) Method

The BCP (bromcresol purple, 5, 5'-dibromo-o-cresolsulfonphtalein) also binds to serum albumin through its sulfophthalein and develops purple color. The advantage of BCP over BCG is that BCP is more specific for albumin than BCG. Therefore the BCP method better agrees with the immunoassay that is the standard method for albumin quantification. On the other hand, the BCP method harbors following problems; a) lower value is shown in patient treated with dialysis because its binding to albumin is inhibited by 3-carboxy-4-methyl-5-propyl-2- furanpropanoic acid (CMPF), an endogenous uremic toxin, accumulating in patients undergoing hemodialysis [13], b) negative error is caused by δ-bilirubin which inhibits the BCP-binding [14], c) the BCG reactivity is different between human albumin and other animal albumin calibrators [15], and d) the BCG reactivity is different among the commercial human serum albumin preparations [4].

Mercaptoalbumin with reduced thiol groups (SH) is oxidized by air to nonmercaptoalbumin forming -S-S- bond. Muramoto points out that the binding reactivity of BCP with albumin is strongly affected by the redox states of thiol (SH) groups in an albumin molecule [4]. Albumin can serve as an antioxidant in blood. The free SH groups are oxidized during reducing oxidants in vivo. Therefore, a ratio of mercaptoalbumin and nonmercaptoalbumin can be varied not only in vivo but also among the purified albumin samples depending on preparation and conservation conditions. Recently, a modified BCP method has been developed, in which effects of the redox states are minimized by adding sodium dodecyl sulphate (SDS) and 5, 5'-[dithiobis] 2-nitrobenzoic acid [4].

MECHANISM OF THE MEASUREMENT INTERFERENCE

Serum albumin binds a variety of endogenous and exogenous substances and plays a role in their transport.

Arvan and his co-wokers suggested that the carboxylic group of the penicillin would increase the negative charge of the molecule, which could account for the observed increase in electrophoretic mobility [8]. Lapresle et al. [16] suggested that the ε-amino group of the albumin covalently binds with a benzylpenicilloyl group which is formed by cleavage of β - lactam ring of penicillin G, increasing electrophoretic migration. Yvon et al. [17] identified the penicillin-binding site of this albumin is the 199th lysine residue. However, it is not proved that the bisalburninemia-like faster electrophoretic mobility caused by penicillin is relevant to the underestimation of albumin in the BCP method. In addition, it is unclear why only the BCP method but neither the BCG method nor the HABA method [18] is affected by penicillin despite they are all dye-binding methods. Sudlow et al. [19] reported five binding sites (I, II, III, metal ion-binding site, and covalent bond site) in albumin for a variety of endogenous and exogenous substances. Lys-190, Lys-199, Lys-432, Lys-541, and Lys-545 in an albumin molecule are reported to be involved in penicillin G-binding [17, 20-23]. It is known that many drugs including warfarin and phenytoin bind to the site I. The site I ranges from ^{199}Lys to ^{292}Glu where amino acids with positive electric charge such as ^{199}Lys and ^{214}His are located in the hydrophobic channeling domain. Reflecting such a structure, there is a report that many drugs which strongly bind to the site I have bulky structure and contain polycyclic aromatic molecule harboring anion charge in the molecular center [24]. Considering that the penicillin structure well adopts this character, the site I may be a good candidate as a penicillin-binding site although the binding site of penicillin is not firmly determined yet. It is very plausible that penicillin shares a same binding site with BCP competitively. Of course, other explanations are possible.

OTHER CLINICAL TESTS INTERFERED BY PENICILLIN

Ppenicillin and its derivatives can interfere with clinical tests other than serum albumin. Penicillin G and ampicillin give positive error in 2, 2′-bicinchoninic acid (BCA) method used for the total protein measurement. Marshall et al. [25] reported that the influence can be avoided by taking the reagent blank. Doumas et al. [26] described that carbenicillin and related antibiotics at a concentration of 0.5 g/L give a positive bias ranging from 0.1 to 1.0 g/L protein in the protein measurement with a biuret reagent. However, they are not expected practically to cause significant interference since therapeutic concentrations of these antibiotics in serum are much lower than 0.5 g/L. δ-Aminolevulinic acid excreted into urine is measured to estimate lead in the body. Ehrlich's aldehyde reagent (*p*-dimethylaminobenzaldehyde in acid solution) is used for detection of δ-aminolevulinic acid. The optical absorbance area of the δ-aminolevulinic acid/Ehrlich's aldehyde reagent complex overlaps that of the complex of penicillin and the reagent, causing false data [27].

CONCLUSION

Penicillin can interfere with various clinical tests. Of them, serum albumin may be most important because this item is one of the most routinely measured clinical tests. In addition, serum albumin measurement by the BCP method is affected by penicillin within a range of therapeutic dose for severe infection, in particular for infective endocarditis. It is also important to know of the albumin measurement system and materials influencing it for proper understanding of the test values of serum albumin.

ACKNOWLEDGMENTS

The effects of antibiotics listed in Table 2 were examined by Wako Pure Chemical Industries (Osaka, Japan). This work was supported in part by Grants-in-Aid for Scientific Research from the Ministry of Education, Science, Technology, Sports, and Culture of Japan.

REFERENCES

[1] Peters TJ. *All about albumin: biochemistry, genetics, and medical applications.* New York: Academic Press, 1995.

[2] Gustafsson JE. Improved specificity of serum albumin determination and estimation of "acute phase reactants" by use of the bromcresol green reaction. *Clin Chem* 1976;22:616-22.

[3] Sogami M, Era S, Nagaoka S, Kuwata K, Kida K, Shigemi J, et al. High-performance liquid chromatographic studies on non-mercapt in equilibrium with mercapt conversion of human serum albumin. II. *J Chromatogr* 1985;332:19-27.

[4] Muramoto Y, Matsushita M, Irino T. Reduction of reaction differences between human mercaptalbumin and human nonmercaptalbumin measured by the bromcresol purple method. *Clin Chim Acta* 1999;289:69-78.

[5] College of American Pathologists. *2010 C-C Participant Summary Reports.* 2010.

[6] Ono M, Aoki Y, Masumoto M, Hotta, T, Uchida Y, Kayamori Y, Kang D. High-dose penicillin G-treatment causes underestimation of serum albumin measured by a modified BCP method. *Clin Chim Acta* 2009; 407: 75.76.

[7] Wilson W, Taubert KA, Gewitz M, Lockhart PB, Baddour LM, Levison M, et al. Prevention of infective endocarditis: guidelines from the American Heart Association: a guideline from the American Heart Association Rheumatic Fever, Endocarditis, and Kawasaki Disease Committee, Council on Cardiovascular Disease in the Young, and the Council on Clinical Cardiology, Council on Cardiovascular Surgery and Anesthesia, and the Quality of Care and Outcomes Research Interdisciplinary Working Group. *Circulation* 2007;116:1736-54.

[8] Arvan DA, Blumberg BS, Melartin L. Transient "bisalbuminemia" induced by drugs. *Clin. Chim. Acta* 1968; 22: 5211-218.

[9] Rodkey FL: Direct spectrophotometric determination of albumin in human serum. *Clin Chem* 1965; 11: 478-487.

[10] Ingwersen S, Raabo E, Improved and more specific bromcresol green methods for the manual and automatic determination of serum albumin. *Clin Chim Acta* 1978; 88: 545-550.

[11] Robertson WS: Optimizing determination of plasma albumin by the bromcresol green dye-binding method *Clin Chem* 1981; 27: 144-146.

[12] Murakami E, Muramoto Y, Kushikamachi A, Shiba K. A reaction rate of serum protein fraction's elutes after electrophoresis to BCG and BCP (in Japanese). *Jap J Med Tech* 1990; 39: 31-35.

[13] Mabuchi H, Nakahashi H: Underestimation of serum albumin by the bromcresol purple method and a major endogenous ligand in uremia. *Clin Chim Acta* 1987; 167: 89-96.

[14] Bush V, Reed RG: Bromcresol purple dye-binding methods underestimate albumin that is carrying covalently bound bilirubin. *Clin Chem* 1987; 33: 821-823.

[15] Assink HA, Blijenberg BG, Boerma GJ, Leijnse B: The introduction of bromocresol purple for the determination of serum albumin on SMAC and ACA, and the standardization procedure. *J. Clin. Chem. & Clin. Biochem* 1984; 22, 685-692.

[16] Lapresle C, Wal JM. The binding of penicillin to albumin molecules in bisalbuminemia induced by penicillin therapy. *Bioch Biophys Acta* 1979; 586: 106-111.

[17] Yvon M, Wal JM. Identification of lysine residue 199 of human serum albumin as a binding site for benzylpenicilloyl groups. *FEBS Lett* 1988;239:237-40.

[18] Arvan DA, Ritz A. Measurement of serum albumin by the HABA-dye technique : A study of the effect of free and conjugated bilirubin, of bile acids and of certain drugs. *Clin Chim Acta* 1969; 26: 505-516.

[19] Sudlow G, Birkett DJ, Wade DN. Further characterization of specific drug binding sites on human serum albumin. Mol *Pharmacol* 1976;12:1052-61.

[20] Lafaye P, Lapresle C. Location of penicilloyl groups on CNBr fragments of the albumin from penicillin-treated patients. *FEBS Lett* 1987;220:206-8.

[21] Lapresle C, Lafaye P. Enzyme-linked immunosorbent assay for measurement of penicilloyl groups. *Ann Inst Pasteur Immunol* 1985;136C:375-82.
[22] Yvon M, Anglade P, Wal JM. Binding of benzyl penicilloyl to human serum albumin. Evidence for a highly reactive region at the junction of domains 1 and 2 of the albumin molecule. *FEBS Lett* 1989;247:273-8.
[23] Yvon M, Anglade P, Wal JM. Identification of the binding sites of benzyl penicilloyl, the allergenic metabolite of penicillin, on the serum albumin molecule. *FEBS Lett* 1990;263:237-40.
[24] Kremer JMH, Wilting J, Janssen LHM. Drug binding to human alpha-1-acid glycoprotein in health and disease. *Parmacol Rev* 1988; 40: 1-47.
[25] Marshall T, Williams KM. Drug interference in the Bradford and 2,2′-bicinchoninic acid protein assays. *Anal Biochem,* 1991; 198: 352-354.
[26] Doumas BT, Bayse DD, Carter RJ, Peters T, Jr, Schaffer R. A Candidate Reference Method for Determination of Total Protein in Serum. I. Development and Validation. *Clin Chem* 1981; 27: 1642-1650.
[27] Kornfelda, JM, Ullmanna WW. Penicillin interference with the determination of δ-aminolevulinic acid. *Clin. Chim. Acta* 1973; 46: 187-190.

In: Penicillin
Editors: G. Barresi and P. Komulainen
ISBN: 978-1-62100-233-8

Chapter 5

Biodegradability Improvement of Penicillin Formulation Effluent by Ozonation and the Fenton Reagents

Idil Arslan-Alaton*
Istanbul Technical University, Faculty of Civil Engineering,
Department of Environmental Engineering, Maslak, Istanbul, Turkey

Abstract

Effects of ozonation and Fenton's treatment on the biodegradability of pharmaceutical wastewater originating from the *Sultamycillin Tosylate Diydrate* penicillin formulation was investigated. Process variables were ozone feed rates (127 – 2750 mg $l^{-1}h^{-1}$), ozonation pH as well as the type of Fenton's reagents (Fenton; Fe^{2+}/H_2O_2 and Fenton-like; Fe^{3+}/H_2O_2) applied at a fixed pH (= 3.0) and different Fe (0.5, 1 mM Fe^{2+} and Fe^{3+}) and H_2O_2 concentrations (10, 20 mM). The efficiency of the selected pre-treatment systems was evaluated in terms of changes in COD (Chemical Oxygen Demand), TOC (Total Organic Carbon) and BOD_5 (5-Day Biochemical Oxygen Demand). The effect of chemical oxidative pre-treatment on activated sludge treatment performance was assessed in terms of microbial oxygen uptake rate (OUR) measurements conducted with unacclimated sewage sludge. Results have demonstrated that

* Fax: + 90 212 285 65 87; Phone: + 90 212 285 65 76; E-mail: arslanid@itu.edu.tr.

although similar overall COD (34 - 42 %) and TOC (24 - 36 %) removal efficiencies have been obtained by the ozonation process and the Fenton and Fenton – like reagents, only ozonation and Fenton-like processes were capable of completely eliminating the inhibitory effect of the penicillin formulation effluent on the microbial consortium in sewage sludge. Moreover, the Fenton-like and ozonation processes appeared to slightly enhance COD abatement rates by 4 % and 8 %, respectively, as could also be demonstrated in separate biodegradation experiments that were run with the raw, pre-treated formulation effluent in the same biosludge.

Keywords: Advanced oxidation, activated sludge inhibition, biodegradability improvement, Fenton's reagent, penicillin formulation effluent, pharmaceutical wastewater, ozonation

INTRODUCTION

Bulk pharmaceutical substances consist of structurally complex organic chemical compounds that are produced within several steps under a variety of precise conditions. Most of the substances involved in the chemical synthesis are listed as priority pollutants [1]. These organic and inorganic compounds that are considered as the principal environmental concern of this sector are generated during the synthesis and formulation steps of production. Hence special attention has recently been devoted to the sustainable, integrated treatment of pharmaceutical formulation effluent [1].

The wide application range of antibiotics in human and veterinary medicine has led to large scale dissemination of refractory and even toxic pollutants in the environment [2,3]. In many countries, a multitude of extremely resistant antibiotics have been found in treated sewage, industrial effluent, the aquatic environment and even in drinking water. These products are discharged into receiving water bodies due to incomplete removal in industrial and municipal treatment plants [4,5]. Most known treatment practices and technologies have appeared to be inappropriate for antibiotic formulation effluent. Hence more advanced treatment technologies such as photochemical (H_2O_2 + UV, O_3 + UV and their combination) and non-photochemical (Fenton's reagent, ozonation at elevated pH, H_2O_2 – assisted ozonation) advanced oxidation processes (AOPs) are required for effective pre-treatment prior to activated sludge process. The main reactant in the above mentioned processes is the hydroxyl radical (•OH) whose oxidation potential

(= 2.8 eV vs. NHE) enables rapid and non-selective oxidation of a variety of inorganic and organic pollutants [6]. A huge body of scientific literature has already been devoted to integrated chemical and biochemical treatment of industrial pollutants such as detergents, phenolic compounds and dyes in textile industry wastewater, chlorinated pollutants from pulp and paper effluents, tanning and bleaching chemicals in tannery effluents to remove recalcitrance and/or toxicity from wastewater [7-20]. Among the studied AOPs, the applied pre-oxidation steps did not always lead to a significant biodegradability enhancement [12, 19,20]. Moreover, photochemically – induced free radical reactions appeared to be less effective than the non-photochemical processes due to high inner UV filter effect of most industrial effluents requiring extreme doses of oxidant and extended reaction periods.

The main purpose of the present work was to demonstrate the pre-treatment efficiency of three different AOPs, i.e. alkaline ozonation (pH = 11.5), Fe(II) and Fe(III) - catalyzed H_2O_2 decomposition at acidic pH (pH = 3.0) on effluent from the penicillin formulation stage of a pharmaceutical industry. The changes in biodegradability of penicillin formulation effluent subjected to optimized ozonation and Fenton processes were studied in more detail. To meet the need for a means of assessing inhibition of ready biodegradability, a combination of BOD_5, microbial sludge oxygen uptake rate and biological degradation experiments was investigated. BOD_5 measurements rather served to screen the optimum pre-treatment period for subsequent biodegradation and respirometric inhibition experiments.

MATERIALS AND METHOD

Penicillin Formulation Effluent

The penicillin effluent used in this study was wastewater from the ALFASID® formulation process (active penicillin ingredient: *Sultamycillin tosylate dihydrate*; Chemical formula: $C_{25}H_{30}N_4O_9S_2.2H_2O$; Molecular weight: 802.85 gr mol^{-1}). The formulation effluent samples were supplied by a pharmaceutical company located in Istanbul, Turkey. The selected effluent corresponded to approximately 30 % of the total daily effluent (= 150 m^3/d), of which approximately 50 % was process water. The collected effluent was stored in plastic carboys at 2 - 4 °C before use for up to four weeks. Prior to chemical and biological experiments, effluent suspensions were filtered

Table 1. Ingredients of the penicillin formulation ALFASID®

Formulation chemical*	Function in the drug formulation
Sultamycillin tosylate dihydrate	active agent
Explotap, dried	additive
Methocel E – 15	additive
Silicone dioksit, colloidal	additive
Magnesium stearate	additive
Microcrystalline cellulose, dried	additive

* The exact amount of each chemical in the final drug formulation mixture is strictly confidential.

through glass fibre filters with a pore size of 1.2 μm to obtain a clear reaction solution and to prevent interference with ozonation and Fenton's treatment. The formulation composition and corresponding effluent characterization (i.e. the mixture of all wash waters) were summarized in Tables 1 and 2, respectively.

All other chemicals used in the study, i.e. oxidant H_2O_2, catalysts $FeSO_4.7H_2O$ and $Fe(NO_3)_3.9H_2O$, concentrated H_2SO_4 and NaOH for pH adjustments, were at least reagent grade.

Ozonation Experiments

1 litre-penicillin wastewater samples were ozonated for 1 hour in a borosilicate glass bubble column in semi-batch mode wherein the ozone + oxygen gas mixture was continuously sparged at a rate of 1.2 l min^{-1} through a fritted gas dispersion disc with a diameter of 5 cm. Ozone was produced by a corona discharge PCI GL-1 model pilot scale ozone generator with a maximum capacity of 20 g h^{-1}. Teflon tubing was used for all connections from the ozone generator to the reaction vessel. All excess (unreacted) gaseous ozone leaving the column was collected in two gas washing bottles connected in series and filled with 10% KI solution, whereas two other gas washing bottles with 2% KI solution were directly placed after the gas introduction line. This was done in order to determine ozone input ($O_{3,in}$) and off-gas ($O_{3,out}$) rates. The ozone feed rate was fixed at 2760 mg l^{-1} h^{-1} for all ozonation experiments. The ozone transfer efficiencies (i.e. absorbed ozone, O_{3A}, in % or mg l^{-1}) were calculated by determining $O_{3,in}$ (ozone input, in mg l^{-1}) and $O_{3,out}$

Table 2. Environmental characterization of the penicillin formulation effluent

Parameter	Unit	Value
Total COD*	mg l^{-1}	687
Soluble COD**	mg l^{-1}	643
TOC	mg l^{-1}	199
BOD_5	mg l^{-1}	16
TKN	mg l^{-1}	85
TP	mg l^{-1}	11
Surfactants	mg l^{-1}	25
Cl^-	mg l^{-1}	95
pH	-	6.85
Total alkalinity	mg $CaCO_3$ l^{-1}	55
Colour***	Cm^{-1}	0.010

* Filtered through 1.20 μm cutoff filter.

** Filtered through 0.45 μm cutoff filter.

*** Measured at wavelength $\lambda = 436$ nm in 1 cm glass cuvettes.

(ozone off-gas, in mg l^{-1}) values for each ozonation experiment using a procedure described by the IOA Standardization Committee [21]. O_{3A} values were calculated as follows;

$$O_{3A}\ (\text{in mg l}^{-1}) = (O_{3,in} - O_{3,out})/O_{3,in} \qquad (1)$$

$$O_{3A}\ (\text{in \%}) = (O_{3,in} - O_{3,out}) \times 100/O_{3,in} \qquad (2)$$

The mass transfer coefficient, k_{La}, of ozone in the semi-batch reactor was determined in acidic pure water by employing the indigo spectrophotometric method [21] as 0.85 ± 0.2 min^{-1}. The same method was also used for the determination of residual, liquid phase ozone concentrations ($O_{3,l}$, in mg l^{-1}) in the reaction solution.

Fenton Experiments

Fenton and Fenton – like experiments were conducted using an ordinary jar test apparatus provided by Velp Instruments, Inc. For that purpose, 0.5 mM and 1 mM $FeSO_4.7H_2O$ or $Fe(NO_3)_3.9H_2O$ were added to 1 litre pre-filtered and H_2SO_4 - acidified (pH = 3) samples as the ferrous and ferric ion catalyst sources, respectively. Thereafter, 10 mM or 20 mM H_2O_2 was introduced from a stock solution (30 % w/w, Merck) that counted as t = 0, i.e. the start of the oxidation reaction. 10 ml – sample aliquots were taken at regular time intervals for 1 hour, however, the reaction proceeded very fast and was complete (leveling off) 20 min after initiation of the reaction. Hence, only final values of the investigated environmental sum parameters have been presented in the present study.

ANALYTICAL PROCEDURE

Raw and chemically pre-treated penicillin formulation effluent samples were immediately analyzed for COD, TOC and BOD_5 according to procedures described in Standard Methods [22]. To prevent the positive interference of H_2O_2 and Fe(II) – iron in the COD measurements, residual (unreacted) H_2O_2 was destroyed with MnO_2 powder at pH = 11, at which pH iron precipitated out as ferric iron hydroxide. The TOC content of the effluent was measured in 0.45 μm filtered air sparged, acidified samples by a Shimadzu TC - 1000 model organic carbon analyzer. The organic carbon analyzer was calibrated with potassium hydrogen phthalate solutions prepared in distilled de-ionized water at varying concentrations.

Biodegradability Experiments

Sewage sludge could not be acclimated to untreated penicillin formulation effluent even after a period of nine weeks corresponding to ten sludge ages. The unassimilated sludge samples used in the biodegradability tests were collected from a laboratory scale 10 liter – capacity municipal wastewater reactor. The mixed culture inoculum was previously fed daily with synthetic sewage that was prepared and stored according to the modified ISO Standard Ref. Nr. 8192 [23]. The ideal mixed liquor volatile suspended solids (MLVSS) concentration that is being typically used in the ISO 8192 activated sludge inhibition test is 2000 - 4000 mg l^{-1} and has been selected as 3000 $\pm$ 150 mg l^{-1} for the present study. The respirometric inhibition tests were performed in 1000 cm^3 capacity glass beakers in which aeration + mixing was provided by means of air diffusion (flow rate = 50 l h^{-1}) and magnetic stirring. The synthetic domestic sewage : penicillin formulation effluent samples were mixed at a ratio of 4 : 1 that resulted in an initial COD of 450 mg l^{-1}. The F/M ratio was selected as 0.15 mg COD (mg MLVSS)$^{-1}$ d^{-1} and the pH was adjusted to 6.9 $\pm$ 0.2 prior to the biodegradability experiments. For the respirometric tests, activated sludge that was periodically fed with the synthetic domestic wastewater and had an initial OUR of 90-110 mg l^{-1} h^{-1}, was used. After initiation of the respirometric experiments oxygen uptake rates (OUR values, in mg l^{-1} h^{-1}) were determined at regular time intervals for the domestic wastewater mixed with raw and chemically pre-treated formulation effluent, using a WTW Oxi 600 model, calibrated oxygen probe. Activated sludge inhibition experiments were conducted for 24 h, i.e. the time when steady state COD values were reached for all tested effluent samples. At the same time, COD abatement was followed for all raw and chemically pre-treated effluent samples mixed with synthetic domestic wastewater. Due to the fact that the sludge could not be acclimated to penicillin formulation effluent under the selected reaction conditions, it was expected that only the synthetic municipal wastewater fraction (75 % of total mixed pharmaceutical wastewater) could be biodegraded under the studied treatment conditions. Accordingly, per cent biochemical COD removal inhibition rate could be determined by the following formula;

$$I_R \text{ (in \%)} = (\Delta COD_{th} - \Delta COD_{exp}) \times 100 / (\Delta COD_{th}) \quad (3)$$

Where I_R is the cent relative inhibition of biological degradation rate of the synthetic municipal wastewater fraction in the effluent sample (in %), $\Delta COD_{th,}$ ΔCOD_{exp} are the theoretically expected (ΔCOD_{th}), and experimentally observed (ΔCOD_{exp}) absolute COD removal rates, respectively (in mg l^{-1}). The above expression shows that the COD fraction which cannot be removed during biodegradation of synthetic sewage in the presence of raw and chemically pre-treated penicillin formulation effluent is positively proportional to the "inhibitory" COD.

Results and Discussion

Pretreatment with Ozone

Figure 1 displays changes in COD and TOC values during ozonation of penicillin formulation effluent at elevated pH. Ozonation experiments have shown that significant COD and TOC removals could only be obtained at elevated pH values (at pH $\geq$11.0) as a consequence of enhanced ozone decomposition into free radicals such as •OH [24]. It is believed that at elevated pH the major oxidizing agent for the degradation of formulation effluent was •OH, since no residual ozone was detectable at pH $\geq$ 7.5 in the reaction medium ($C_{O3, l}$ was $\leq$ 0.3 mg/L). No attempts have been made to control the pH of the formulation effluent during ozonation, that gradually decreased to pH = 7.3 – 7.5 at the end of the reaction period. Consequently no further pH adjustment was required prior to BOD_5 measurements, respirometric inhibition and biodegradation experiments.

At the end of the ozonation period 233 mg l^{-1} (34 %) COD was removed and 742 mg l^{-1} of the applied O_3 was absorbed in the reaction solution. TOC abatement rates remained rather low (24 % overall reduction) throughout the ozonation period, however, TOC removal was actually not the purpose of the present work.

Upon closer inspection of Figure 2 presenting changes in ozone absorption rates (O_{3A} values) for an ozonation experiment run at pH = 11.0, it can be concluded that a linear relationship exists between ozone feed rates and ozone absorption efficiencies. However, only 30 % of the introduced ozone could be absorbed throughout the ozonation period. Percent ozone absorption efficiency was high at the beginning of the reaction (i.e. the first 10 min of ozonation), because the pH was high and hence ozone decomposition was significantly enhanced during this period. Upon extended ozonation the reaction pH decreased to neutral values (pH = 7.0 - 7.5) due to acid accumulation, so that ozone decomposition was slightly retarded. This was also evidenced by increases in liquid phase, residual ozone concentrations when the pH dropped to below 7 ($O_{3,l}$ = 1.1. mg l^{-1}). The above indicated findings suggest that the acceleration in ozone decomposition is mainly governed by the reaction pH and not the COD removal efficiency, i.e. ozone consumption due to enhanced oxidation. Therefore no direct correlation between oxidation efficiency and ozone decomposition seemed plausible.

The rate of partial oxidation of the penicillin formulation ingredients to biologically more degradable compounds relative to the rate of mineralization to oxidation end products such as carbonate, inorganic salts and H_2O, can be expressed as the COD TOC^{-1} ratio. Changes in the COD, TOC values observed during ozonation of penicillin formulation effluent has already been shown in Figure 1. The "partial oxidation ratio" can be derived from the COD, TOC data given in Figure 1 for alkaline ozonation. It could be concluded that the COD TOC^{-1} ratio fell down from 3.5 at t = 0 to 3.0 after t = 1 h, indicating that though COD and TOC abatements more or less paralleled each other, partial oxidation (COD abatement) rates were still appreciably higher than mineralization (TOC abatement) rates during ozonation of penicillin formulation effluent.

Figure 3 presents the changes in the biodegradability of penicillin formulation effluent, expressed as the BOD_5 values and the BOD_5/COD ratio, as a function of ozonation time. From Figure 3 it is evident that the biodegradability of the formulation effluent was improved from only 0.02 for the original, untreated penicillin formulation wastewater (original BOD_5 = 17 $\pm$ 2) to 0.12 at the end of the ozonation period reaching a maximum value of 0.27 (BOD_5= 128 mg l^{-1}) after an ozonation period of 40 min corresponding to an ozone feed rate of 1670 mg l^{-1}. Obtained results revealed that even though non-acclimated sludge has been used for the BOD_5 tests, a relative increase in BOD_5 values was apparent for wastewater samples ozonated at a dose of 1670 mg l^{-1}, indicating that ozonation at a pollutant - specific, "optimum" dose may have positive effects on the biological treatability of the otherwise refractory wastewater. The existence of a distinct, maximum BOD_5/COD ratio and BOD_5 value at an optimum ozone dose has already been evidenced in previous studies [11,12,19,20]. As a general rule, it has been established that an ozone dose of at least 1-2 mg O_3 mg COD_o^{-1} or 2-3 mg O_3 mg TOC_o^{-1} is practically required to achieve the highest possible biodegradability improvement.

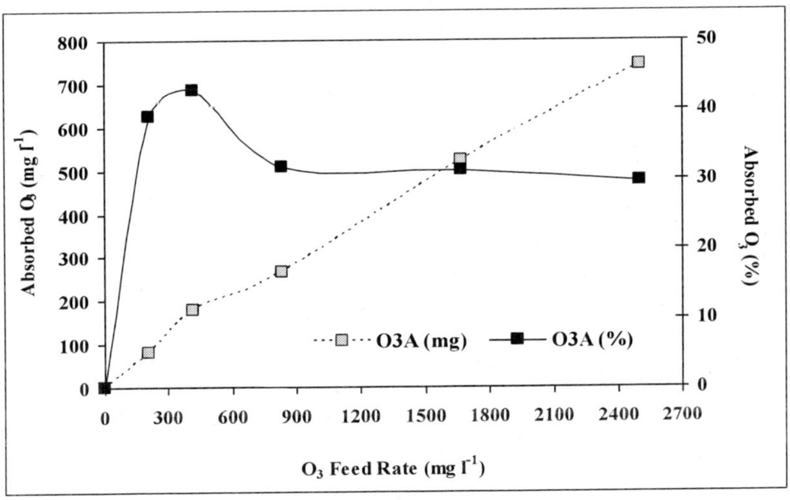

Figure 1. Changes in COD and TOC removal rates observed during ozonation of penicillin formulation effluent at pH_o = 11.0.

Pretreatment with Fenton and Fenton-Like Processes

In previous studies, Fenton and Fenton-like reactions were found as promising candiates for effective pre-oxidation and reduction in biorecalcitrance [25]. Consequently, the Fenton process and its modification, i.e. the Fenton-like reaction, were studied as the chemical pre-treatment alternatives to ozonation. As can be seen in Figure 4, similar COD removal rates were found by applying the Fenton and Fenton-like processes as compared with ozonation at pH=11.5. Fenton oxidation systems were studied at different Fe – iron (ferrous or ferric iron salts) : H_2O_2 molar concentration ratios in accordance with molar ratios reported in the previous, scientific literature [9, 25-27]. Obviously, the partial oxidation efficiency increased at higher catalyst and oxidant doses. Again, partial oxidation rates were accelerated upon increasing the oxidant and catalyst dose or switching from the Fenton to Fenton-like processes.

It is well known that H_2O_2 decomposition into free radical species can be achived via both ferrous and ferric iron catalysts [25,26]. The apparently higher overall COD and TOC removal rates found with the Fenton-like reagent as compared with the conventional Fenton's reaction is due to the fact that two different removal mechanisms are attributable for the abatement efficiency in the latter case; i.e. oxidation via free radicals formed as a consequence of H_2O_2 decomposition at acidic pH plus removal via accompanied ferric iron coagulation observed after pH re-adjustment. Fenton's coagulation or Fenton co-precipitation occurs with the in-situ formed ferric iron hydroxide at pH > 5, 6 [25]. In conclusion, the enhancement was attributable to the fact that Fenton's oxidaiton plus co-precipitation is enhanced in the presence of ferric rather than ferrous iron.

However, BOD_5 values and BOD_5/COD ratios first appeared to increase upon increasing the H_2O_2 concentraiton from 10 mM to 20 mM in the presence of 0.5 mM $Fe^{2+/3+}$ and to decrease upon further addition of Fe-catalyst to a concentration of 1 mM. As in the case of ozonation, optimum chemical pre-oxidation conditions had to be defined and over-oxidation to be prevented in order to improve the biotreatability of the penicillin formulation effluent. For instance, the biodegradability (BOD_5/COD) ratio was accelerated to 0.30 and 0.36 for the Fenton and Fenton-like processes conducted with 20 mM H_2O_2, respectively, when the $Fe^{2+/3+}$: H_2O_2 ratio was kept at 1:40 instead of 1:20 at 1 mM $Fe^{2+/3+}$, at which dose the obtained BOD_5/COD pre-oxidation ratio fell down to 0.26-0.28 due to over-oxidation at higher catalyst doses.

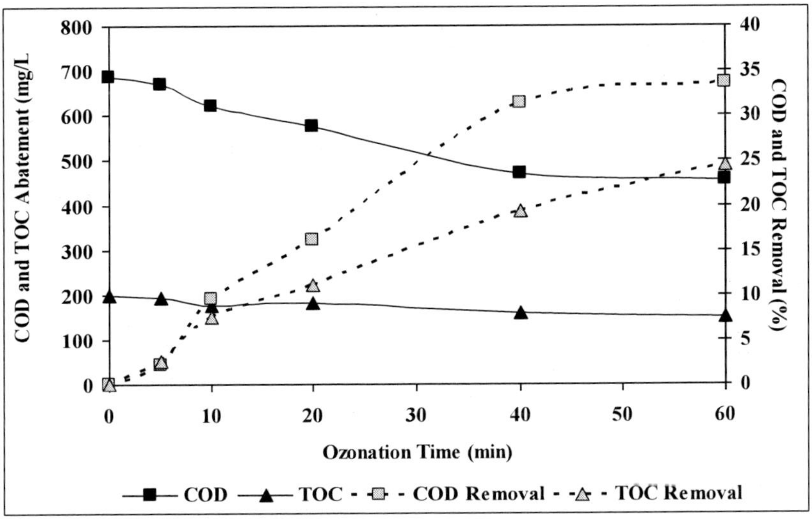

Figure 2. Changes in ozone absorption rates (O_{3A} values) observed during ozonation of penicillin formulation effluent at pH_o = 11.0 with respect to ozone feed rate.

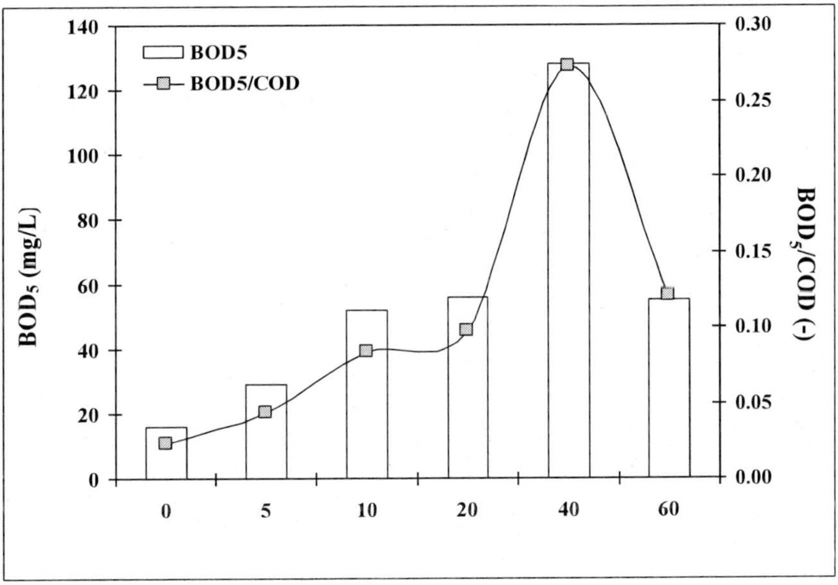

Figure 3. Changes in BOD_5 and BOD_5/COD values observed during ozonation of penicillin formulation effluent at pH_o = 11.0.

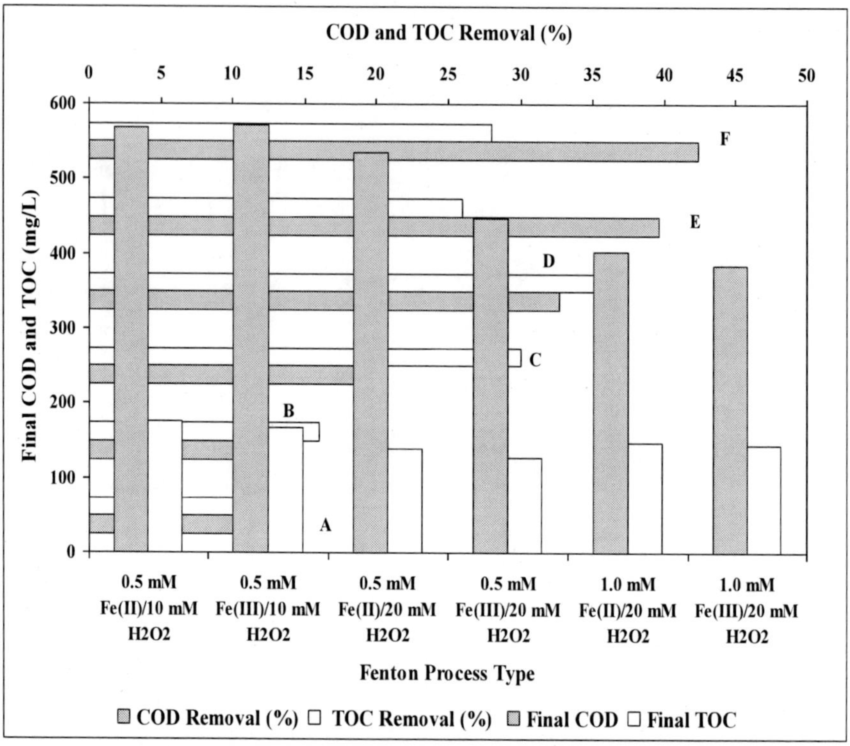

Figure 4. Percent COD and TOC removal efficiencies and final COD and TOC values obtained after 25 min Fenton and Fenton-like treatment at pH = 3. The Fen+ concentration /type and H2O2 concentration are given in the figure and indicated as condition A (0.5 mM Fe2+ + 10 mM H2O2), B (0.5 mM Fe3+ + 10 mM H2O2), C (0.5 mM Fe2+ + 20 mM H2O2), D (0.5 mM Fe3+ + 20 mM H2O2), E (1 mM Fe2+ + 20 mM H2O2) and F (1 mM Fe3+ + 20 mM H2O2).

Though the evaluation of relative changes in the BOD_5/COD ratio is most commonly practised in studies dealing with oxidative pre-treatment of refractory industrial wastewater via chemical oxidation processes, it is only an approximate tool and not sufficient to predict the fate of industrial effluents in real activated sludge systems. In other words, the above obtained experimental data is not very informative and has to be supported by actual biodegradation studies as to prove that chemical pre-oxidation with ozone and Fenton's reagent is a well suited, sustainable approach for penicillin formulation effluent.

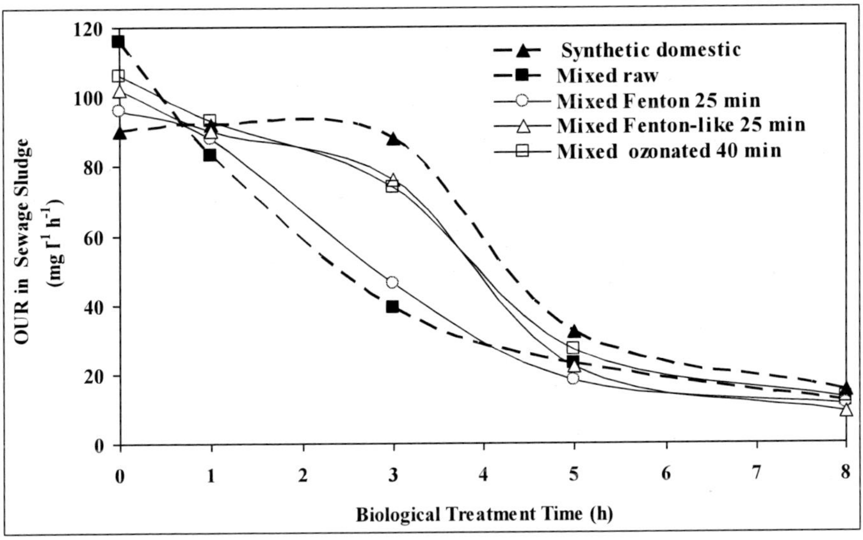

Figure 5. Oxygen uptake rates (OUR values, in mg l^{-1} h^{-1}) measured for unacclimated sewage sludge during biological treatment of raw, ozonated (at $pH_o = 11.0$), Fenton and Fenton-like pre-treated penicillin formulation effluent + synthetic domestic wastewater mixture. Experimental conditions are given in the text.

Activated Sludge Inhibition Tests

Oxygen uptake rate measurements have often been used as an appropriate tool to long-term and shock mimic responses of industrial pollutants in activated sludge treatment works [28-30]. Figure 5 depicts oxygen uptake rates (OUR values, in mg l^{-1} h^{-1}) obtained during 8 h batch biodegradation tests conducted with raw (untreated) and chemically pre-treated penicillin formulation effluent (30% v/v) + synthetic wastewater (70% v/v) mixtures. In addition, Table 3 presents percent COD inhibition rates (I_R values) that have been determined for the successive biodegradation of raw and chemically pre-treated wastewater samples. Upon careful comparison of Figure it is quite obvious that the data given in Figure 5 and Table 3 support each other. For instance, Figure 5 demonstrates that effluent samples which were previously treated with ozone at alkaline pH or with the Fenton – like reagent (Fe^{3+}/H_2O_2) almost behaved like synthetic domestic effluent in terms of respirometric activity. Table 3 on the other hand shows that for the mixed raw wastewater biochemical COD removal was inhibited by 27 % relatively to the that of synthetic domestic effluent. The rate of COD inhibition was reduced from 27

% to 16 %, when the Fenton process was previously applied to the penicillin effluent fraction, however, was still a positive value and hence inhibition existed for Fenton-treated wastewater. In the case of alkaline ozonation and Fenton-like processes, complete reduction in the originally inert COD fraction of the penicillin formulation effluent was observed, thus resulting in "negative" inhibition rates, i.e. a rather positive effect of pre-ozonation and the Fenton-like reagent was obtained for the latter two pre-treatment systems by 4 % and 8 %, respectively. Conclusively, not only the synthetic domestic wastewater, but also the formulation effluent fraction was degradable upon preliminary treatment with ozonation and Fenton-like reagent.

From the above findings it can be inferred that although according to the BOD_5/COD approach, all three selected advanced oxidation systems were practically equally effective in biodegradability enhancement (provided that the AOPs were first optimized for oxidant and catalyst dose), the OUR activity and COD biodegradation tests confirmed that activated sludge inhibition can be reduced, but not completely hindered by Fenton's pre-treatment, but with applying the Fenton-like process and in particular alkaline ozonation. On the other hand ozonation appears to be the technically and economically more feasible oxidation method since only one pH - adjustment would be necessary prior to ozonation to accelerate ozone decomposition thereby initiating free radical reactions.

Table 3. Biological COD removal rates and percent I_R obtained for raw and chemically pre-treated penicillin formulation effluent

Treatment Type	Final COD (mg l^{-1})	Biological COD removal (%)	I_R (%)
No pre-treatment	475	47	+ 33
Fenton (pH = 3.0)*	369	59	+ 16
Fenton-like (pH = 3.0)*	235	73	- 4 ***
Ozonation (pH = 11.0)**	193	76	- 8 ***

* H_2O_2 : Fe molar ratio = 40 (0.5 mM Fe and 20 mM H_2O_2).

** Ozone feed rate = 42 mg min^{-1} (applied ozone dose = 1680 mg l^{-1} or 1.9 mg O_3 mg COD_o^{-1}).

*** The negative sign ("-") indicates that biodegradation is not inhibited, but enhanced.

Experimental Conditions: Ozonation at high pH (= 11.0); Fenton and Fenton-like treatment at acidic pH (= 3.0)

CONCLUSION

The aim of the present experimental study was to evaluate the changes in biocompatibility of penicillin formulation effluent, being selected as a refractory index wastewater, after oxidative chemical pre-treatment by ozonation and the Fenton's reagents. Chemical oxidative treatment performance of the selected advanced oxidation processes was evaluated in terms of COD and TOC abatement rates. Changes in biodegradability have been assessed in terms of the conventional BOD_5 parameter that has been determined for each process, the electrolytic respirometric inhibition test and successive biological COD abatement experiments.

For that purpose, in addition to BOD_5 measurements, exogenous oxygen uptake rates were followed for activated sludge being exposed to a mixture of synthetic domestic wastewater and raw, chemically pre-treated penicillin formulation effluent at fixed food-to-microorganism ratios. Moreover, several biodegradation experiments were performed to question the validity and relationship of the selected methodological approaches.

It was demonstrated that all investigated procedures are capable of reflecting the situation in activated sludge systems. However, the evaluation of exogenous activated sludge oxygen uptake and COD abatement rates which were followed during successive biodegradation experiments were both required to get a deeper insight. In the particular case it could be established that chemical pre-oxidation of penicillin formulation effluent may completely reduce if not minimize its inhibition towards sewage sludge in aerobic treatment systems, provided that the chemical oxidant type, its application conditions (concentration/dose, treatment time, pH etc.) were specifically optimized for the industrial effluent in question.

In conclusion, chemical oxidation of penicillin formulation effluent prior to its discharge into biological treatment units is not only expected to eliminate its negative impacts on microbial activity of the sewer system but also to facilitate its more economic and safe degradation via conventional, cost effective treatment methods.

ACKNOWLEDGMENTS

The author wishes to thank TUBITAK (Turkish National Scientific and Technology Research Council) for the financial support under Project "ICTAG –C-042".

REFERENCES

[1] EU Direction General III, *Assessment of potential risks to the environment posed by medical products for human use (excluding products containing live genetically modified organisms)*. No. 5504/94 Draft 6, Version 4, 5th January, European Union, Brussels (1995).

[2] Kümmerer, K., Steger-Hartmann, T. and Meyer, M., Biodegradability of the anti- tumor agent ifosfamide and its occurence in hospital effluents and communal sewage. *Wat. Res.*, 31(11), 2705-2710 (1997).

[3] Tisler, T. and Zagorc-Koncan, J., Toxicity evaluation of wastewater from the pharmaceutical industry to aquatic organisms. *Wat. Sci. Tech.*, 39(10-11), 71-76 (1999).

[4] Wollenberger, L., Halling-Sorensen, B. and Kusk, K. O. Acute and chronic toxicity of veterinary antibiotics to *Dapnia magna*. *Chemospher*e, 40, 723-730 (2000).

[5] Kümmerer, K., Al-Ahmad, A. and Mersch-Sundermann, V., Biodegradability of some antibiotics, elimination of the genotoxicity and affection of wastewater bacteria in a simple test. *Chemosphere*, 40, 701-710 (2000).

[6] Oppenlaender, T. *Photochemical purification of water and air*. Wiley VCH Verlag, Weinheim, pp. 5-19 (2003).

[7] Zwiener, C. and Frimmel, F. H., Oxidative treatment of pharmaceuticals in water. *Wat. Res.*, 34(6), 1881-1885 (2000).

[8] Gulyas, H., von Bismarck, R. and Hemmerling, L., Treatment of industrial wastewaters with ozone/hydrogen peroxide. *Wat. Sci. Tech.*, 32(7), 127-134 (1995).

[9] Koyama, O., Kamagata, Y. and Nakanura, K., Degradation of chlorinated aromatics by Fenton oxidation and methanogenic digester sludge. *Wat. Res.*, 28, 885-899 (1994).

[10] Saupe, A. and Wiesman, U., Ozonization of 2,4-dinitrotoluene and 4-nitroaniline as well as improved dissolved organic carbon removal by

sequential ozonization – biodegradation. *Wat. Environ. Res.*, 70(2), 146-154 (1998).

[11] Benitez, F. J., Acero, J. L., Gonzalez, T. and Garcia, J., Ozonation and biodegradation processes in batch reactors treating black table olives washing wastewaters. *Ind. Eng. Chem. Res.*, 40, 3144-3251 (2001).

[12] Alvares A. B. C., Diaper, C., Parsons, S. A., Partial oxidation by ozone to remove recalcitrance from wastewaters - a review. *Environ. Technol.*, 22(4), 409-427 (2001).

[13] Arslan-Alaton, I., Kornmüller, A. and Jekel, M. R., Ozonation of a spent reactive dye bath: effect of CO_3^{2-}/HCO_3^- alkalinity. *J. Environ. Eng., ASCE*, 128(8), 689-696 (2002).

[14] Sarria, V., Parra, S., Adler, N., Peringer, P., Benitez, N. and Pulgarin, C., Recent developments in the coupling of photoassisted and aerobic biological processes for the treatment of biorecalcitrant compounds. *Catal. Today*, 76, 301-315 (2002).

[15] Esplugas, S., Gimenez, J., Contreras, S., Pascual, E. and Rodriguez, Comparison of different advanced oxidation processes for phenol degradation. *Wat. Res.*, 36, 1034-1042 (2002).

[16] Hess, T. F. and Schrader, P. S., Coupled abiotic – biotic mineralization of 2,4,6,-Trinitrotoluene (TNT). *J. Environ. Qual.*, 31, 736-744 (2002).

[17] Hernandez, R., Zappi, M., Colucci, J. and Jones, R., J., Comparing the performance of various advanced oxidation processes for treatment of acetone contaminated water. *J. Haz. Mat.*, 92, 33-50 (2002).

[18] Perez, M., Torrades, F., Domenech, X. And Peral J., Removal of organic contaminants in paper pulp effluents by AOPs: An economic study. *J. Chem. Technol. Biotechnol.*, 77, 525-531 (2002).

[19] Scott, J. P. and Ollis, D. F., Integration of chemical and biological oxidation processes for water treatment: review and recommendations. *Environ. Prog.*, 14(2), 88-103 (1992).

[20] Arslan-Alaton, I., Akmehmet Balcioglu, I., and Bahnemann, D. W., Advanced oxidation of a reactive dyebath effluent: Comparison of O_3, H_2O_2/UV-C and TiO_2/UV-A processes. *Wat. Res.*, 36, 1143-1154 (2002).

[21] IOA Standardisation Committee - Europe, 001/87-F, Iodometric method for the determination of ozone in a process gas. Brussels (1987).

[22] APHA-AWWA-WEF, *Standard Method for the Examination of Water and Wastewater,* 19th Ed. American Public Health Association, Washington DC (1995).

[23] ISO 8192, Water quality – oxygen demand inhibition assay by activated sludge (1986).
[24] Staehelin, J. and Hoigné, J., Decomposition of ozone in water: Rate of initiation by hydroxide ions and hydrogen peroxide. *Environ. Sci. Technol.* 16(10), 676-681 (1982).
[25] Arslan, I. and Akmehmet Balcioglu, I., Degradation of commercial reactive dyestuffs by heterogeneous and homogeneous advanced oxidation processes: a comparative study. *Dyes Pigments*, 43, 95-108 (1999).
[26] De Laat, J. and Gallard, H., Catalytic decomposition of hydrogen peroxide by Fe(III) in homogenous aqueous solution: Mechanism and kinetic modelling. *Environ. Sci. Technol.*, 33, 2726-2732 (1999).
[27] Haber, F. and Weiss, J., The catalytic decomposition of hydrogen peroxide by iron salts. *Proc. Royal Soc. London, Ser. A*, 147, 332-351 (1934).
[28] Gotvajn, A. Z. and Zagorc-Koncan, J., Biodegradation studies as an important way to estimate the environmental fate of chemicals. *Wat. Sci. Tech.*, 39(10-11), 375-382 (1999).
[29] Orupold, K. Hellat, K. and Tenno, T., Estimation of treatability of different industrial wastewaters by activated sludge oxygen uptake measurements. *Wat. Sci. Tech.*, 40(1), 31-36 (1999).
[30] Guiterrez, M.; Extebarria, J. de las Fuentas, L., Evaluation of wastewater toxicity: comparative study between Microtox and activated sludge oxygen uptake inhibition. *Wat. Res.*, 36, 919-924 (2002).

In: Penicillin
Editors: G. Barresi and P. Komulainen
ISBN: 978-1-62100-233-8

Chapter 6

GEOMETRICAL AND ELECTRONIC STRUCTURE OF MOLECULE BENZILPENICILLIN BY METHOD AB INITIO

***V. A. Babkin*[1]*, V. Yu. Dmitriev*[1] *and G. E. Zaikov*[2]**

[1]Volgograd State Architect-Build University Sebrykov Department, Michailovka, Volgograd Region, Russia

[2]Institute of Biochemical Physics Russian Academy of Sciences, Moscow, Russia

ABSTRACT

Quantum chemical calculation of a molecule of benzilpenicillin is executed by method AB INITIO in base 6-311G** with optimization of geometry on all parameters for the first time. The optimized geometrical and electronic structure of this compound is received. Acid force of benzilpenicillin is theoretically appreciated. It is established, than it to relate to a class of weak H-acids (pKa=+12, where pKa-universal index of acidity).

Keywords: quantum chemical calculation, method AB INITIO, benzilpenitsilin, acid power

Aims and Background

The Aim of this work is a study of electronic structure of molecule benzilpenicillin and theoretical estimation its acid power by quantum-chemical method AB INITIO in base 6-311G**. The calculation was done with optimization of all parameters by standard gradient method built-in in PC GAMESS [1]. The calculation was executed in approach the insulated molecule in gas phase. Program MacMolPlt was used for visual presentation of the model of the molecule. [2].

Methodical Part

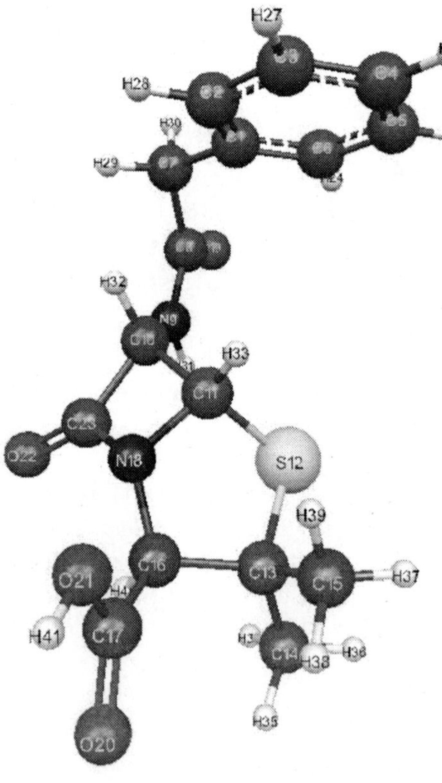

Figure 1. Geometric and electronic molecule structure of benzilpenicillin. (E_0= -3728247 kDg/mol, E_{el}= -9375191 kDg/mol).

Geometric and electronic structures, general and electronic energies of molecule benzilpenicillin was received by method AB INITIO in base 6-311G** and are shown on Figure and in table The universal factor of acidity was calculated by formula: pKa = 49,4-134,61*q_{max}^{H+} [3] (where, q_{max}^{H+} – a maximum positive charge on atom of the hydrogen q_{max}^{H+}= +0.28 (for benzilpenicillin q_{max}^{H+} alike table)). pKa=12.

Quantum-chemical calculation of molecule benzilpenicillin by method AB INITIO in base 6-311G** was executed for the first time. Optimized geometric and electronic structures of thise compound was received. Acid power of molecule benzilpenicillin was theoretically evaluated (pKa=12). This compound pertain to class of weak H-acids (9<pKa<14).

Table 1. Optimized bond lengths, valence corners and charges on atoms of the molecule of benzilpenicillin

Bond lengths	R,A	Valence corners	Grad	Atom	Charge (by Milliken)
C(1)-C(2)	1.38			C1	-0.20
C(2)-C(3)	1.38	C(3)C(2)C(1)	120	C2	-0.09
C(3)-C(4)	1.38	C(4)C(3)C(2)	120	C3	-0.09
C(4)-C(5)	1.38	C(5)C(4)C(3)	119	C4	-0.10
C(5)-C(6)	1.51	C(6)C(5)C(4)	120	C5	-0.08
C(1)-C(7)	1.52	C(7)C(1)C(6)	120	C6	-0.06
C(7)-C(8)	1.36	C(8)C(7)C(1)	113	C7	-0.16
C(8)-N(9)	1.41	N(9)C(8)C(7)	118	C8	0.51
N(9)-C(10)	1.55	C(10)C(9)C(8)	128	N9	-0.49
C(10)-C(11)	1.55	C(11)C(10)C(9)	121	C10	-0.03
C(11)-S(12)	1.85	S(12)C(11)C(10)	117	C11	-0.07
S(12)-C(13)	1.52	C(13)S(12)C(11)	90	S12	0.12
C(13)-C(14)	1.53	C(14)C(13)S(12)	109	C13	-0.46
C(13)-C(15)	1.57	C(15)C(13)S(12)	109	C14	-0.18
C(13)-C(16)	1.51	C(16)C(13)S(12)	103	C15	-0.16
C(16)-C(17)	1.44	C(17)C(16)C(13)	111	C16	0.07
C(16)-N(18)	1.3	N(18)C(16)C(13)	107	C17	0.54
C(8)-O(19)	1.3	O(19)C(8)C(7)	121	N18	-0.41
C(17)-O(20)	1.31	O(20)C(17)C(16)	122	O19	-0.47
C(17)-O(21)	1.32	O(21)C(17)C(16)	114	O20	-0.45
N(18)-O(22)	1.3	O(22)N(18)C(16)	102	O21	-0.38
O(22)-C(23)	1.07	C(23)O(22)N(18)	24	O22	-0.43
C(6)-H(24)	1.07	H(24)C(6)C(5)	119	C23	0.53

Table 1. (Continued)

Bond lengths	R,A	Valence corners	Grad	Atom	Charge (by Milliken)
C(5)-H(25)	1.07	H(25)C(5)C(6)	119	H24	0.12
C(4)-H(26)	1.07	H(26)C(4)C(3)	120	H25	0.10
C(3)-H(27)	1.07	H(27)C(3)C(2)	119	H26	0.10
C(2)-H(28)	1.07	H(28)C(2)C(3)	119	H27	0.10
C(7)-H(29)	1.08	H(29)C(7)C(1)	110	H28	0.09
C(7)-H(30)	1.08	H(30)C(7)H(29)	107	H29	0.12
N(9)-H(31)	0.99	H(31)N(9)C(8)	113	H30	0.16
C(10)-H(32)	1.07	H(32)C(10)N(9)	111	H31	0.26
C(11)-H(33)	1.07	H(33)C(11)N(18)	115	H32	0.16
C(14)-H(34)	1.08	H(34)C(14)C(13)	111	H33	0.17
C(14)-H(35)	1.08	H(35)C(14)H(34)	108	H34	0.11
C(14)-H(36)	1.08	H(36)C(14)H(35)	108	H35	0.13
C(15)-H(37)	1.08	H(37)C(15)C(13)	109	H36	0.11
C(15)-H(38)	1.08	H(38)C(15)H(37)	108	H37	0.12
C(15)-H(39)	1.08	H(39)C(15)H(37)	108	H38	0.11
C(16)-H(40)	1.08	H(40)C(16)N(18)	108	H39	0.11
O(21)-H(41)	0.94	H(41)O(21)C(17)	108	H40	0.18
				H41	0.28

References

[1] Shmidt, M.W., Baldrosge, K.K., Elbert, J.A., Gordon, M.S., Enseh, J.H., Koseki S., Matsvnaga. N., Nguyen K.A., Su S.J., and others. *J. Comput. Chem*. 14, 1347-1363, (1993).

[2] Babkin V.A., Fedunov R.G., Minsker K.S. and others. *Oxidation Communication*, 2002, №1, 25, 21-47.

[3] Bode, B. M. and Gordon, M. S. *J. Mol. Graphics Mod*., 16, 1998, 133-138.

INDEX

#

21st century, 60

A

abatement, xiii, 150, 156, 157, 158, 160, 165, 166
Abraham, 7, 31, 32, 33, 48, 49, 50, 56, 58, 61, 66
absorption, 157, 161
access, 122
accessibility, 124
acetic acid, 36
acetone, 168
acetylation, 65
acid, x, 2, 5, 7, 8, 12, 14, 15, 19, 20, 21, 23, 25, 27, 28, 30, 35, 40, 45, 47, 49, 50, 52, 57, 58, 60, 63, 66, 73, 77, 95, 111, 112, 118, 121, 136, 142, 143, 144, 148, 158, 171, 172
acidic, 12, 112, 140, 151, 154, 160, 165
acidity, xiv, 171, 173
active site, 32, 33, 37, 38, 41, 50, 58, 117, 118, 120, 121, 122, 124
acylation, 129
adjustment, xi, 70, 157, 160, 164
adsorption, 73
advancement, 5, 25
adverse effects, ix
aerobic, 166, 168
agar, 3, 36
alanine, 43, 122
albumin, xiii, 134, 135, 136, 137, 138, 139, 140, 142, 143, 144, 145, 146, 147, 148
alcohols, 54, 74
alimentary canal, 13
alkaline, 151, 158, 164
alkalinity, 154, 168
American Heart Association, xii, 133, 146
amine, 72, 76, 77, 78, 79, 82, 111
amines, 73, 74, 76
amino, xii, 7, 13, 16, 19, 25, 28, 32, 33, 40, 43, 44, 45, 47, 51, 55, 112, 113, 116, 121, 125, 134, 144
amino acid, 7, 19, 25, 28, 32, 33, 40, 43, 44, 45, 47, 51, 55, 112, 113, 125, 134, 144
amino acids, 7, 19, 44, 47, 112, 113, 125, 134, 144
amino groups, 44
anti-, 167
antibiotic, ix, xii, 2, 3, 5, 6, 17, 18, 19, 23, 26, 36, 48, 49, 50, 52, 59, 63, 65, 66, 72, 116, 117, 127, 139, 151
antibiotic resistance, 6, 116, 127
antibiotics, 150, 167
antibody, xii, 115, 118, 119, 120, 126, 127, 128, 129, 131
antigen, xii, 116, 118, 121
antioxidant, 143
APA, 5, 8, 13, 15, 23, 27, 63

aqueous solutions, xi, 70, 93
arginine, 43, 44, 56
aromatics, 167
ascorbic acid, 35
aspartate, 32, 59
aspartic acid, 32
atoms, 173
ATP, 35, 52

B

Bacillus subtilis, 127
bacteria, ix, xii, 1, 3, 4, 6, 12, 13, 22, 23, 34, 115, 117, 119, 124, 127, 167
bacterial infection, xii, 133
bacteriophage, 119, 122, 130
base, xiv, 56, 118, 129, 171, 172, 173
Belgium, 125, 126
beneficial effect, 48
bias, 145
bile, 147
bile acids, 147
bilirubin, 134, 143, 147
bioassay, 36, 40
biochemistry, 32, 146
biocompatibility, 165
bioconversion, 48, 55
biodegradability, xiii, 149, 150, 151, 152, 155, 158, 160, 164, 165
biodegradation, xiii, 150, 152, 156, 157, 162, 163, 164, 165, 166, 168
biological activities, 61
biological processes, 168
biomarkers, 119
biomass, 75, 80, 81, 87, 91, 92, 94, 108
biomolecules, xii, 115, 117, 119, 125, 127
biosensors, 119
biosynthesis, ix, 1, 5, 7, 12, 17, 18, 31, 34, 47, 49, 50, 52, 53, 54, 59, 60, 61, 62, 65
biosynthetic pathways, ix, 1, 5, 7, 8, 31
biotic, 168
biuret, 137, 145
bleaching, 151
blood, 14, 134, 139, 142, 143
bonding, 63
branching, 7
butyl ether, 79
by-products, 38

C

cancer, 119
candidates, xii, 116
CAP, 135
carbon, 13, 38, 47, 155, 167
carbon dioxide, 38
carbon molecule, 47
carboxyl, 16
catalysis, 31, 32, 38, 39, 46, 51, 61, 118
catalyst, 118, 122, 154, 159, 160, 164
catalytic activity, xii, 34, 38, 42, 43, 45, 51, 116, 123, 130
catalytic properties, 46
C-C, 146
cellulose, 36, 152
cephalosporin, ix, 1, 2, 3, 4, 5, 7, 8, 9, 11, 12, 16, 21, 22, 23, 25, 26, 27, 28, 29, 30, 31, 36, 37, 46, 47, 48, 49, 50, 51, 52, 54, 55, 56, 57, 58, 59, 60, 61, 62, 63, 65, 67
cephalosporin antibiotics, 3, 5, 51, 60
ceramic, 110
challenges, 27
chemical, x, xiii, xiv, 2, 5, 7, 14, 16, 21, 24, 25, 32, 34, 56, 90, 108, 150, 151, 152, 159, 160, 162, 165, 166, 168, 171, 172, 173
chemicals, xi, 70, 151, 153, 169
chloroform, 79
chromatography, 36, 40
chronic, 167
classes, 4, 46
classification, v
cleavage, 144
clinical application, 21
clinical trials, 53
clone, 40, 42
cloning, 6, 32, 59, 60
closure, 31, 32
coagulation, 160
coding, x, 2, 119

coenzyme, 8, 50, 66
color, 142, 143
commercial, 143, 169
composition, 75, 80, 153
compounds, xii, 3, 5, 6, 36, 48, 56, 58, 73, 116, 150, 158, 168
concentration, 155, 159, 160, 162, 166
concentration ratios, 159
condensation, 12
configuration, 27, 95
conjugated bilirubin, 147
conservation, 143
construction, 39, 50
consumption, 73, 158
contaminant, 3
contaminated water, 168
contamination, 29, 75
coordination, 8, 37
copper, 113
corona discharge, 153
correlation, 67, 82, 101, 103, 104, 105, 106, 158
correlations, 101, 104
cost, x, 46, 69, 73, 83, 90, 108, 166
covalent bond, 144
covering, 75
crystal structure, 31, 32, 37, 39, 42, 44, 128
crystallisation, 37
crystals, 38
culture, 8, 12, 13, 155
cysteine, 7, 33, 59, 121
cystic fibrosis, 16, 67

D

D-amino acids, 33
deaths, 6
decomposition, 73, 79, 151, 157, 160, 164, 169
defence, 117, 124
degradation, 4, 25, 125, 152, 156, 157, 166, 168
degradation rate, 156
dendritic cell (DC), 169
derivatives, xii, 8, 21, 28, 117, 123, 133, 134, 144
detectable, 35, 157
detection, 35, 36, 145
detergents, 151
dialysis, 73, 143
diffusion, 73, 79, 87, 98, 100, 105, 156
diffusivity, 71, 104
diluent, 74, 78, 80, 82, 89, 90
dimethylsulfoxide, 126
discrimination, 120
diseases, 3, 5, 6, 12, 46, 48
dispersion, 75, 153
dissociation, 79
distribution, xi, 70, 74, 76, 80, 81, 82, 86, 87, 89, 92, 98, 100, 101, 103, 107, 108, 110, 111, 125
diversity, 117
DNA, 41, 60, 119, 122, 130, 131
DNA polymerase, 41, 131
donors, 31
dosage, 18
drinking water, 151
drug design, 125, 127
drug discovery, 127, 129
drug targets, 47
drugs, 50, 119, 127, 134, 135, 139, 141, 144, 146, 147
dyes, 141, 142, 151

E

effluent, xiii, 150, 151, 152, 153, 155, 156, 157, 158, 159, 160, 161, 162, 163, 164, 165, 166, 168
effluents, 151, 162, 167, 168
electric charge, 144
electron, 31
electronic structure, xiv, 171, 172, 173
electrophoresis, xiii, 133, 135, 136, 137, 139, 147
ELISA, 123
elucidation, ix, 1, 5, 38
emulsions, 76

encoding, 18, 19, 20, 27, 28, 29, 52, 53, 56, 60, 119
endocarditis, xii, 133, 135, 137, 139, 145, 146
energy, 75, 124
engineering, ix, 2, 40, 42, 119, 130
ent, 165
environment, 44, 104, 151, 166
environmental impact, 90, 108
enzymatic activity, 37, 39
enzyme, x, 2, 8, 9, 10, 11, 27, 32, 33, 35, 36, 37, 38, 39, 40, 41, 42, 44, 45, 47, 51, 53, 56, 57, 65, 67, 123, 126, 128
enzymes, ix, 2, 6, 9, 11, 12, 18, 20, 31, 35, 43, 46, 47, 50, 55, 56, 57, 59, 60, 62, 63, 117, 121, 127
equilibrium, 37, 71, 77, 79, 80, 92, 98, 107, 110, 112, 146
equipment, 101
ethanol, 18, 20
EU, 166
Europe, 169
European Union, 167
evidence, 37, 53, 56
evolution, 48, 55, 62, 66, 119, 127
exogenous, 165, 166
experimental condition, 76, 101
exploitation, 34
exposure, 56
extraction, xi, 70, 71, 73, 74, 75, 76, 78, 79, 80, 81, 82, 84, 85, 86, 87, 89, 91, 92, 93, 94, 95, 97, 103, 107, 108, 109, 110, 111, 112, 113
extracts, 51, 59, 73

F

fermentation, x, 2, 5, 6, 8, 12, 13, 25, 29, 34, 50, 61, 65, 69, 72, 74, 75, 80, 81, 86, 87, 105, 106, 108, 110
ferric ion, 154
fiber, vii, ix, 70, 71, 74, 88, 95, 102, 104, 106, 112, 113
fibers, 71, 88, 95, 98, 102, 104, 107
fidelity, 41
films, 78
filters, 153
filtration, 37
financial, 109, 166
financial support, 166
first generation, 22
flexibility, 38, 44, 61, 74, 123
flotation, 110
fluorescence, 54
food, 165
force, xiv, 171
formation, x, 2, 7, 12, 19, 20, 31, 32, 33, 36, 37, 48, 62, 65, 74, 76, 82
formula, 152, 156, 173
fragments, 148
France, 115
free radicals, 157, 160
fungi, ix, 1, 3, 6, 8, 19, 62, 65
fungus, ix, 1
furan, 23
fusion, 7, 47

G

gel, 37, 135
genes, ix, 1, 6, 9, 11, 12, 17, 18, 19, 20, 27, 28, 30, 40, 50, 52, 53, 56, 57, 59, 64, 65, 121
genetic linkage, 7
genetics, 58, 146
genome, 17, 53, 119
genus, 4
geometry, xiv, 171
graph, 86
growth, 36, 54, 128
growth factor, 128
guidelines, 146

H

harbors, 143
health, 108, 148
heart disease, 129
heme, 67
hemodialysis, 143

hepatitis, 131
heptane, 79
hexane, 74
histidine, 32, 64
homogenous, 169
host, 18, 32, 39
human, 116, 128, 130, 143, 146, 147, 148, 150, 167
human immunodeficiency virus, 130
hybrid, 12, 39, 40, 48, 121
hydrocarbons, 74
hydrogen, 25, 33, 71, 77, 79, 155, 167, 169, 173
hydrogen peroxide, 25, 167, 169
hydrolysis, 23, 120, 126, 128
hydrophobicity, 44
hydroxide, 155, 160, 169
hydroxyl, 9, 13, 21, 151
hypothesis, 38, 117, 118

I

ideal, 25, 26, 44, 45, 155
identification, 32
image, xii, 116, 118
immune system, 117, 128
immunocompromised, 16
immunoelectrophoresis, 142
immunoglobulin, 118
improvements, 45
in-, 160
in vitro, 22, 33, 35, 40
in vivo, 27, 28, 29, 37, 39, 48, 134, 143
incubation time, xiii, 134
industries, 13
industry, 151, 152, 167
infection, 145
ingredients, 158
inhibition, 3, 36, 40, 41, 42, 66, 116, 121, 125, 126, 128, 150, 152, 155, 156, 157, 163, 164, 165, 166, 169, 170
inhibitor, 14, 23, 50, 60, 121, 124
initiation, 155, 156, 169
inoculum, 155
inorganic, 150, 151, 158
insertion, 119
integrins, 128
integrity, 44
intensive care unit, 66
interface, 72, 76, 78, 92, 97, 98, 101, 110, 111, 130, 131
interference, 141, 145, 148, 153, 155
intravenously, 136
ions, 41, 82, 134, 169
iron, 8, 31, 33, 35, 37, 38, 63, 67, 155, 159, 160, 169
isolation, 17, 49, 54
isoleucine, 44
isozymes, 51

J

Japan, 133, 137, 142, 145

K

kerosene, x, 69, 74, 78, 82, 89, 104, 110
kinetic, 169
kinetic model, 169
kinetic studies, 36, 123
kinetics, 32, 75, 76, 78

L

laminar, 104
lead, 19, 21, 25, 127, 145, 151
leucine, 32, 43, 44, 51
lifetime, 125
ligand, 121, 122, 124, 128, 147
light, 12, 36, 37
light scattering, 37
liquid chromatography, 36, 57
liquid phase, 95, 154, 158
liquids, 109, 113
liver, 134
liver disease, 135
low temperatures, 73
lumen, 95
Luo, 112

M

machinery, 116, 119
malabsorption, 135
malignancy, 135
malnutrition, 135
mammalian cells, 3
manganese, 41
manipulation, x, 2, 6, 7, 19, 26, 28, 118
mapping, 5
mass, 71, 74, 75, 90, 92, 93, 95, 97, 99, 100, 101, 102, 103, 104, 105, 106, 107, 108, 113, 154
materials, 134, 145
matrix, 124
measurement, xiii, 121, 134, 136, 137, 141, 142, 145, 148
measurements, xiii, 80, 123, 139, 150, 152, 155, 157, 163, 165, 169
media, x, 69, 75, 80, 81, 105, 106
medical, 3, 146, 166
medical history, 3
medicine, 150
melanin, 39
membrane separation processes, 110
membranes, 95
metabolism, 119
metabolites, 53, 134
metal ion, 144
metals, 113
methyl group, 9
methyl groups, 9
methylation, 9
mg, xiii, 149, 153, 154, 155, 156, 157, 158, 159, 163, 164, 165
microorganism, 165
microorganisms, ix, 1, 6, 7, 11, 17, 54, 57
migration, 144
mimicry, 127
mineralization, 158, 168
Ministry of Education, 145
mixing, 156
mode, 153
modelling, 93, 110, 111, 124, 128, 169
models, 119, 124
modifications, ix, 16, 17, 63
modules, 12, 47, 53, 63, 90, 95, 104, 112
molar ratios, 159
molecular biology, ix, 1, 6
molecular oxygen, 9, 61
molecular weight, 126, 134
molecules, xii, 33, 78, 82, 97, 104, 107, 116, 117, 118, 124, 147
monoclonal antibody, 130
monomers, 37
Montenegro, 49, 54, 56
Moscow, 171
motif, 8, 31, 119
mutagenesis, 5, 17, 26, 32, 38, 40, 43, 64, 67, 118, 122
mutant, 40, 41, 42, 46
mutation, 41, 43, 44, 45, 48, 56
mutations, x, 2, 41, 43, 45, 55, 58, 60, 67

N

NADH, 10
natural selection, 5
nephrotic syndrome, 135
Netherlands, 17
network theory, 128
neutral, 157
New Zealand, 80, 109
nitrite, 55
non-, 151, 158
nuclei, 47
nucleophiles, 118
nucleotides, 41
nucleus, 5, 9, 15, 16, 23, 25, 26, 28, 34, 38

O

OH, 151, 157
oil, xi, 70, 75, 82, 85, 86, 87, 90, 92, 93, 96, 106, 108
olive oil, 80
operon, 57
optimization, xiv, 62, 127, 171, 172
organic, 150, 151, 155, 167, 168
organic solvents, xi, 24, 69, 89

organism, 39, 61
osmotic pressure, 134
overproduction, 51
oxidation, 35, 150, 151, 155, 158, 159, 160, 162, 164, 165, 166, 167, 168, 169
oxidation rate, 160
oxidative reaction, 34, 37
oxygen, xiii, 31, 150, 152, 153, 156, 163, 165, 166, 169, 170
ozonation, xiii, 149, 150, 151, 153, 154, 157, 158, 159, 160, 161, 164, 165
ozone, xiii, 149, 153, 154, 157, 158, 161, 162, 164, 165, 167, 168, 169
ozonization, 167

P

partition, 80, 89
pathogens, 6, 13, 14, 46, 47, 48, 49
pathways, 7, 49
PCR, 41
penicillium notatum, ix, 1, 3, 62
peptide, 12, 34, 52, 53, 54, 56, 64, 119, 120, 123, 126, 129, 131
peptides, xii, 58, 116, 120, 125, 128, 129, 131
permeation, 109
Petroleum, 69
pH, xi, xiii, 35, 70, 73, 75, 76, 79, 80, 81, 82, 83, 84, 85, 86, 87, 88, 89, 92, 94, 107, 108, 111, 126, 134, 149, 151, 153, 154, 155, 156, 157, 158, 159, 160, 162, 164, 165, 166
phage, xii, 116, 119, 123, 127, 128, 129, 130, 131
pharmaceutical, xiii, 3, 17, 46, 149, 150, 151, 152, 156, 167
pharmaceuticals, x, 69, 167
pharmacokinetics, 47
phenol, 168
phenolic compounds, 151
phenotype, 58
phenylalanine, 44
phenytoin, 144
phosphate, 9, 80, 82, 89, 105, 109, 126
plants, 151
plutonium, 112
point mutation, 41
polar, 44, 74, 80
polarity, 80, 107
pollutants, 150, 163
polymer, 109, 110, 111
polypeptides, 47
polypropylene, 88, 95
pools, 19
porosity, 72, 105
potassium, 155
pre-, xiii, 149, 151, 154, 155, 156, 159, 160, 162, 163, 164, 165, 166
precipitation, 73, 136, 160
preparation, 23, 79, 143
preservation, 54
prevention, 135, 139
probe, 43, 156
process gas, 169
producers, x, 2, 7, 18, 19, 23, 26, 28
prokaryotes, 34
proliferation, 14
proline, 123, 126
promoter, 18, 20
protein design, 119
protein engineering, 25
protein family, 34
protein sequence, 8
proteins, x, xii, 2, 3, 10, 32, 52, 54, 66, 116, 117, 122, 125, 126, 128, 129, 137
proteolysis, xii, 116
Pseudomonas aeruginosa, 55, 58, 60, 64, 66, 67, 130
public health, 127
pulp, 151, 168
pure water, 154
purification, 62, 110, 167

Q

quantification, 140, 143

R

radical reactions, 151, 164
radio, 40
radius, 102
re-, 160
reactant, 63, 151
reactants, 142, 146
reaction mechanism, 74, 77, 95
reaction medium, 157
reaction rate, 147
reaction temperature, 40
reactions, ix, 2, 12, 35, 47, 107, 159
reactivity, 8, 130, 135, 143
reagents, xiii, 38, 149, 165
receptors, 125
recognition, 35, 46, 123, 129
recombinant DNA, 6, 17, 18, 65
recombination, 39, 40, 46, 48
recommendations, 168
recovery, ix, x, 69, 73, 74, 82, 93, 108, 109, 112
recycling, 102
regeneration, 73, 112, 122
relevance, 48
requirements, 60
researchers, 3, 41, 106
residues, xii, 32, 33, 37, 39, 40, 43, 44, 61, 64, 116, 118, 121, 122, 123, 124, 125, 126, 129
resistance, xii, 6, 12, 14, 21, 22, 46, 47, 48, 55, 56, 58, 59, 60, 64, 87, 90, 92, 108, 115, 116, 117, 125, 127, 128
resolution, 43
response, xii, 115, 117, 127
rings, 33
risks, 166
room temperature, 76, 126, 139
Russia, 171

S

safety, 53, 104, 108
salts, 158, 159, 169
saturation, 92
secondary metabolism, 65
secrete, 39
selectivity, 53, 73
semi-, 153, 154
sensitivity, 125
serine, 117
serum, ix, xii, 116, 133, 134, 135, 136, 137, 138, 139, 140, 141, 142, 143, 144, 145, 146, 147, 148
serum albumin, ix, xiii, 133, 135, 136, 137, 139, 140, 141, 142, 143, 144, 145, 146, 147, 148
services, iv
sewage, xiii, 150, 151, 155, 156, 163, 166, 167
shock, 163
showing, 125, 137
side chain, 5, 8, 12, 13, 16, 21, 22, 23, 25, 27, 28, 33, 44, 47
side effects, 125
sign, 165
signals, 123
simulation, 112, 124
Singapore, 1
single chain, 118
skin, 22, 53
sludge, xiii, 150, 151, 152, 155, 158, 162, 163, 164, 165, 166, 167, 169, 170
sodium, xi, 70, 79, 93, 108, 143
software, 122, 124
solid phase, 36
solubility, 36, 73, 76
solution, xi, 38, 70, 71, 72, 79, 80, 81, 82, 83, 84, 87, 88, 89, 91, 92, 94, 97, 101, 102, 103, 104, 105, 107, 108, 110, 111, 122, 145, 153, 154, 157, 169
solvents, x, 69, 74, 75, 79, 80, 82, 83, 89, 90, 93, 107, 108
Spain, 17, 113
species, ix, 1, 4, 33, 38, 62, 125, 160
spectrophotometric method, 154
spectroscopy, 54
stability, 22, 37, 47, 60, 61, 111
standardization, 147
staphylococci, 12, 13, 16, 57, 65

state, 37, 71, 73, 82, 89, 95, 101, 156
states, 135, 143
steady state, 156
stoichiometry, 78, 79, 95, 107
strain improvement, 17, 58
streptococci, 16
structural characteristics, 34
structural gene, 18, 20
structure, 3, 5, 7, 13, 14, 21, 22, 31, 32, 34, 37, 39, 42, 44, 45, 47, 48, 51, 53, 58, 60, 62, 117, 118, 121, 142, 144, 173
substitution, 33, 41, 45
substitutions, 40, 45
substrate, x, 2, 8, 25, 27, 28, 31, 32, 33, 34, 35, 36, 37, 38, 39, 40, 41, 42, 43, 44, 45, 46, 47, 49, 51, 56, 57, 59, 61, 62, 67, 128, 129
substrates, 10, 31, 34, 35, 38, 43, 44, 47
suicide, 128
sulphur, 3, 4, 38
supported liquid membrane, 95, 109, 112, 113
surface area, 70, 95
surfactant, 111
suspensions, 152
swelling, 74, 104
synthesis, x, xii, 2, 3, 5, 8, 12, 21, 23, 32, 34, 47, 59, 63, 115, 116, 125, 127, 150

T

Taiwan, 17
tandem repeats, 54
target, xii, 5, 47, 104, 116, 119, 123, 130
taxonomy, 61
TBP, xi, 69
techniques, ix, 1, 6, 17, 65, 73
technologies, 111, 151
technology, 6, 112, 119, 127
TEM, xii, 116, 120, 121, 122, 123, 124, 125, 126, 129
temperature, xi, 35, 70, 73, 75
testing, 47
therapeutic agents, xii, 116, 119
therapeutic use, v
therapeutics, 116
therapy, xiii, 51, 63, 134, 147
time frame, 93
toluene, 74
toxic waste, 25
toxicity, 3, 6, 72, 74, 83, 90, 107, 134, 151, 167, 169
toxin, 143
transfer, 153, 154
transformation, x, 2
transport, 54, 65, 97, 98, 107, 144
treatment, xii, xiii, 3, 5, 22, 46, 53, 133, 138, 146, 149, 150, 151, 153, 156, 159, 162, 163, 164, 165, 166, 167, 168
treatment methods, 166
trial, 23
tributyl phosphate, xi, 69, 74, 80, 82, 89, 104, 107
trojan horses, 117
tryptophan, 44, 123, 124, 126
tumor, 167
Turkey, 149, 152
turnover, 46
tyrosine, 44

U

UK, 17
United, 22, 51, 55, 69, 109
urea, 16
urine, 145
USA, 16, 80, 88, 95, 122, 129
UV, 80, 151, 169

V

valence, 173
valine, 7, 8, 31, 33, 53, 56, 57, 64
variables, xiii, 75, 149
vector, 40
vegetable oil, xi, 69, 82
velocity, 71, 102, 103, 104, 106
vessel, 153
viruses, 119
viscosity, 72, 74, 89

W

Washington, 128, 169
waste, 73, 113
waste water, 113
wastewater, xiii, 149, 150, 151, 152, 153, 155, 156, 158, 162, 163, 165, 167, 169
water, 31, 36, 76, 77, 96, 112, 113, 151, 152, 155, 167, 168, 169
weakness, xii, 116
web, 130
Wisconsin, 17, 18, 20
workers, 18, 28, 35, 36, 37, 43
World Health Organisation, 6, 109
worldwide, 6

X

X-ray diffraction, 50

Y

yield, 5, 6, 7, 9, 18, 19, 20, 23, 25, 26, 27, 28, 36, 73